CI GRAPHICS

IN JAPAN

CI GRAPHICS IN JAPAN

RIKUYO-SHA

CONTENTS

目次

Preface

Yusaku Kamekura
CI Graphic Editor in chief, Graphic Designer

The development of corporate identity —CI— as a management strategy was introduced to the Japanese business world only five or six years ago, and in that short space of time an astonishing number of companies have undergone this baptism-like adoption of names and development of images through typographic and graphic design. The concept of CI is usually thought of as originating in the United States, but in fact the seminal example of CI strategy was provided by the Italian company Olivetti. Olivetti became the center of attention world wide when it began to take a cultural approach to the designs used to express its corporate image. At that time even in the United States, corporations did not dream that design could have a profound impact on business performance. Once America's largest corporation, IBM, became aware of Olivetti's use of design as a management strategy, however, CI quickly became the focus of attention in the U.S. business world. In time, almost every American company was caught up in the CI whirlwind, but it was about ten years until its effects began to be felt among Japanese corporations.

Up until that time, we Japanese had interpreted the design activities at IBM and Olivetti strictly in terms of design policy; we did not understand that the concept of CI had attained the status of a highly developed theory of management. Today, CI is as much in fashion among Japanese enterprises as it was once in the United States. Corporations are using CI strategies to help them survive when they fall on hard times, to reinvigorate their business or to provide the "shot-in-the-arm" needed to carry them forward to new spurts of growth. Some rather simplistically believe that all they need is a new and effective CI to inject new life into their internal corporate structure. Of course, acquiring a genuine and distinctive CI is not that simple.

I do not wish to launch into a full-fleged discussion of the concept of CI here. As you will see, this book does not deal in-depth with the nature of CI. As a book, it can at best give us but a relatively superficial view of the graphic expressions of CI. The editors have been aware of and troubled by this dilemma from the inception of the idea for the book. We cannot avoid the inherent contradictions posed in trying to present as many works as possible while seeking to present each in the most skillful way possible. CI in its true form is many-faceted and takes concrete shape only after years of cautious and incisive study. If we tried to capture that real image of CI in book form, it would not doubt require three or four volumes to do justice to, for example, that of IBM alone.

We realized the danger that, no matter how we might contrive, it might appear to be no more than a kind of digest of the visual aspects of CI. We could have simply called it a catalog of trademarks and corporate symbols, but a book with such a simplistic concept is unlikely see the light of day in the publishing world.

I was the one who chose the title "CI Graphics," although I do not even know whether this is a term that might be called "natural" in English. Even if it is not, I would like the reader to recognize it as a word coined by Japanese. I chose this title because I realized that the book deals only with the visual dimension of CI. It deals, in fact, with CI in which design plays the main role. No matter how one describes it, this is a book conceived and edited by designers, and this genesis is responsible both for its merits and its faults.

We have tried to collect the examples from the most outstanding CI plans and to select them on the basis of visual excellence, aesthetic qualities, and accuracy in expressing the corporate image.

We have sought to be as conscientious as possible within the limits of the realities we had to work with, and we would like to ask the reader's indulgence on this account.

序

亀倉雄策　CIグラフィックス編集委員長、グラフィックデザイナー

CIが経営戦略として、日本の企業界に登場して、それがこの5、6年の間に驚くべき多くの会社が、その洗礼を受けた。ご存知のようにCIの発祥の地はアメリカだと思う。しかしアメリカのCIに多くのヒントを与えたのは、イタリアのオリベッティ社である。オリベッティは、企業表現としてデザインを文化という面で捉えて世界中の注目をあびた。その頃、まだアメリカの企業はデザインが経営に大きな影響を与えるとは夢にも思っていなかった。しかしやがてアメリカ最大の企業IBMが、くしくもこのオリベッティの経営とデザインの関係に気づいたことから、アメリカのCI戦略が、次第に熱をおびてきたといわれている。やがてアメリカの企業界には、CI旋風が吹きまくるわけだが、それから10数年もたって、やっと日本にCIらしい芽がふき出した。
それまで私達はオリベッティやIBMのデザイン活動を、デザインポリシーという解釈をしていて、CIという深い経営理念にまで到達する思想までは、理解出来なかった。それが、現在の日本企業においては、一時期のアメリカを彷彿させるほどのCIの大流行になったのである。CIによって企業は生残りをはかり、さらに企業を活性化し、そうして次の飛躍のためのエネルギーにしようとしている。それは、CIを採用さえすれば、企業内部に活力が生まれると安易に信じているからである。しかしCIというのは、そんな簡単なものではないのだ。
私は、今ここでCI論を正面から論ずるつもりはない。この本をごらんになるとおわかりのように、この本ではCIの深い重要な部分にふれてはいない。本という性質上、CIという顔を手のひらでなでてみた程度でしか、ふれることが出来なかった。この悩みは編集はもちろん、発想の段階からいだきつづけたものだ。一冊の本のなかに、いかに多くの作例を、いかに上手に構成してみせるかが本の生命という点では、どうしても矛盾を感じてしまう。CIは底が深くて、長い年月にわたってたえず、注意深く時代を鋭く読みとり、具体的に実行に移していくのが本当の姿である。しかし、もしこの真実の姿を本にまとめるということになれば、例えばIBMだけでも三冊や四冊の本になってしまうだろう。だから矛盾を充分承知の上で本をつくらねばならず、編集のむずかしさを、つくづく感じないわけにはいかない。要するに、どんなに工夫をこらしても、CIの視覚面のダイジェストになってしまうからである。だからCIといわずにトレードマークとシンボルといってしまえば簡単なことだが、そういった単純なテーマの設定では、今ではもう出版するチャンスがない。出版事業というものの限界を考えたうえで、私達編集委員が、どこまで良識を生かせるかが、問題点であったということを理解していただきたい。そのためには出来る限り、優れたCIの実施例を集めて、それに視覚的な完成度、感性、企業表現の正確さまでを考慮して選択しなくてはならないのは当然である。
「CI GRAPHICS」というタイトルは、私が考えたものだが、このタイトルが英語としてまた言葉として世界に通用するかどうかわからない。日本人が考えた造語として了解してほしい。CIにGRAPHICSをつけたのは、本の性質上、視覚面だけを捉えた内容になることを充分に承知してつけたつもりである。その意味では、デザインを主役として考えたCIだということも出来よう。ちらしにしてもデザイナーが考え、デザイナーが編集した本である点によさもあるし、悪さもあることは否定出来ないと思う。

Corporate Identity in the High-tech Society

Hiroshi Kashiwagi

Assistant Professor, Tokyo University of Art and Design (Tokyo Zokei Daigaku)

Just when "corporate identity" (CI)—the concept as it was developed in Europe and the United States—took root in Japan is difficult to say precisely. It was not until the 1960s that professional design journals started to carry case studies introducing the total CI design strategies of such corporations as Italy's Olivetti, West Germany's Braun, and America's IBM and Westinghouse. Certainly well before that time Japanese corporations and other organizations had some knowledge of and interest in the concept of corporate identity. They knew that logos and symbolic marks were an important dimension of a corporation's image. Before the war, the most popular professional journal of advertising and graphic design of the time, *Koukokukai* (published by Seibundo Shinkosha) put out a special issue in October 1936 presenting a collection of cuts of Olympic events. It had been planned that the 1940 Olympic Games would be held in Tokyo (this plan was, of course, terminated by the war), and the collection was part of the total design project for the Games. But the concept of corporate identity had yet to take root at this time.

It was not until the 1970s that the expression "corporate identity" (borrowing the English term) really began to spread among Japanese corporations and other organizations as well as among graphic designers. In other words, even before that time, design work for the purposes of CI strategy was going on, but it was not until the seventies that the term itself was used to describe it. As the concept of "CI" became established, there was a remarkable increase in the number of organizations—public, private, and corporate—that were utilizing symbolic marks, logos, and other graphic means of promoting a unified image.

Since the beginning of the 1980s, CI strategy has become almost an obsession in this country. Corporations and other organizations find it a full-time task to create new and better symbols and typo-graphic logos to improve their images and keep them up-to-date, a job that even involves, from time to time, redesigning their buildings and other space facilities. CI has never before been such a central concern of organizations nor so widely put into practice in corporate strategy.

The CI Boom and Technological Change

Why has CI been such an obsession in Japan since the beginning of the 1980s? Perhaps the most direct reason is the need faced by corporations and public organizations to project stronger images toward the consumer and the public. If they do not, they are doomed to remain shadowy, undefined presences, lost in the flood of images and information that sweeps through the market today. Another direct reason is that from the beginning of the 1980s, consumer trends and tastes have began to change very rapidly and corporations and public organizations had to act quickly to refurbish their images in keeping with the times. It is not surprising that CI has been discussed and put into practice to a degree that would seem almost excessive.

Why did trends begin to change so rapidly at the beginning of the eighties? It was because the age of machine technology had given way to the age of electronics technology. Amid the flood of corporate images and information, companies found themselves pressed to formulate corporate identities that had clarity and impact. They also had to reestablish their corporate images strategically, taking into account the new tastes and trends of the times. What it all came down to, however, was the necessity to cope with the transition from the machine age to the age of electronics.

Technology is not a cold and detached element of our existence. It is very closely related to both the market economy and culture. Our society is thoroughly governed by the logic of the market economy (i.e., capital), and this logic dominates not only the realm of economic activity, but of culture as well. The logic of the market economy, furthermore, is intricately linked to the development of technology.

There are many examples in the history of design of the way the interrelationships between technology, the market economy, and culture determine the sense of the times and the designs which give that sense concrete form. For example, the period between World War I and World War II was a time when machine technology became a part of everyday life. Of course, the industrial revolution had occurred long before, in the latter part of the nineteenth century, but it was not until the inter-war period in the early twentieth century that machines became a common feature of daily life.

Machine technology deals only with the functional aspect of different elements and serves to rationalize, or economize, as much as possible the relationship between those elements. Machine technology, therefore, is well suited to the logic of economy. It made it possible to produce things cheaply and in vast quantities. Modern society is characterized by the emergence of the mass consumer. Mass consumption, of course, would not be possible without machine technology which supplied the market with inexpensive goods produced in vast quantities. Under the logic of machine technology, everything from graphics to architectural space was expressed in designs that were above all functionalist and based on the principles of mechanical rationalism (i.e., efficiency first). Machine technology was a means of efficiency and economy. It associated itself with the "masses" by facilitating mass production and mass consumption, and one of its features was standardized design.

Following reconstruction in the wake of the defeat in World War II, Japan entered a period of rapid economic growth beginning in the later fifties and extending throughout the sixties. During that period the mass consumption society became firmly entrenched, a result, it hardly needs saying, of machine technology. Today, mechanical technology is rapidly being replaced by electronics technology, and it may not be long before mechanical technology is largely obsolete. Of course, it is unlikely that machines will ever pass completely out of existence, but their technology, at the very least, will no longer reign supreme. As it happens, it was in the early part of the 1980s that the mechanical technology that had once been ascendant began to go into obsolencence. This period of technological revolution was paralleled by the surge of interest in corporate identity in Japan, and the coincidence is not purely accidental.

Eroding Identities

About two years ago, a journalist from abroad approached me (since I have made the study of corporate identity something of a specialty) with several questions about design in Japan. One of his questions revealed that he had quite a good grasp of the characteristics of Japanese design from the beginning of the 1980s. He observed that the designs of household appliances made by West Germany's Braun and Holland's Philips, for example, were so distinctive that you could tell which company had made which product, whether it be a piece of audio equipment or a shaver. The appliances made by particular Japanese makers, he said, did not have such unified, easily recognizable images. He said he thought that the proliferation of designs seemed only to be escalating.

He was right. The designs of oven toasters made by a single manufacturer have no unifying characteristic. They include pink and emerald green pop-art models as well as sleek and stylish Hi-Tec (which is, in the lingo of the design world, an abbreviation for "High Style" [stylish] and High Technology) versions. Naturally, the designs of a company's products contribute nothing to the creation of readily recognizable image. And this trend does seem to be growing stronger, as the journalist pointed out.

The diversification of product designs is a phenomenon resulting from the "multiple types/small quantities" mode of production, as it is described in the marketing field. Of course, this diversification is theoretically a problem separate from that of the older problem of inconsistency in product design. As I observed earlier, mass production and mass consumption were built on machine technology. Those forms of production and consumption gave rise to standardized designs. T-shirts, for example, were mass-produced by the millions. When goods like this entered the market, everyone could afford a T-shirt, and, of course, everyone looked alike.

In time, however, people began to feel a desperate urge to escape from the suffocating uniformity and standardization that forced everyone to look the same. That urge grew ever more manifest as the multiple-types/small quantity mode of production made it possible for a far greater variety of products than ever seen before to enter the market. Our latent desire for variety was brought to the surface and further stimulated by the appearance of diversified merchandise on the market.

A typical example of the multiple-types/small-quantity mode of production are the American "Cabbage-Patch" dolls, a case which economist Katsuhito Iwai discusses in his book *Benisu no shōnin no shihonron* [Capitalism According to the Merchant of Venice] (Chikuma Shobō, 1985). What made this mode of production possible is electronics technology, pre-eminently the computer. It is hard to tell whether the trend will grow even more pronounced or not, but it is fairly certain that electronics technology will make possible even further diversity achieved by increasing the number of elements (parts) from which products are made and their interchangeability.

As electronics technology becomes part of our daily lives, a wide variety of designs catering to the tastes of consumers has appeared. For the manufacturer busily churning out products in as great variety as possible, it is practically impossible to hope for designs that might endow the company with a distinctive image. This being the case, it is inevitable that corporations' images tend to erode unless constant effort is made to combat trend.

Design, needless to say, deals with "signs" or symbols, and today it involves far more than symbolism alone, but embraces "information" as well. In short, design is information, and, because of this it is expected to be fresh and immediate. As a result, new designs appear one after another in a never-ending stream of diverse and self-asserting forms. In this flood of information and designs, the images of corporations and public organizations are bound to erode. This is the reason for the strong demand we see today for development of CI strategies aimed at building corporate symbols and well-integrated images.

The workings of the economy, if left to follow pure principle, naturally move in the direction of infinite expansion. Theoretically, then, the dynamics of the economy brush aside all obstacles to its expansion, be they religious or ethical constraints or barriers of social custom. In the absence of human intervention, an economy possesses a self-generating power. The self-expanding nature of economic activity can be explained as a process of "decodage."

Human society, however, does not allow economic activity to go on unconstrained. We are incessantly reorganizing it, fencing it about with various rules, and trying to recodify it. In other words, the development of an economy involves a constant tug-of-war between the forces that seek to overstep the established codes and the forces striving to hem them in and reorganize them. Those codes, meanwhile, are always changing.

Today, as a result of electronics technology, the workings of the economy are obviously breaking free of national boundaries. Is it possible that capital will eventually transcend all constraints of nationality or territory? Perhaps, considering the way multinational corporations have grown and the emergence of an expanding network of information that links every part of the world. It is unrealistic, however, to think that capital will shed nationality and national boundaries completely. Inevitably, powerful forces will emerge that will try to impose constraints on that trend.

In any case, Japan is being transformed by high technology and corporations are being internationalized as they leap national boundaries. When capital and information transcend boundaries it means, naturally, that people are crossing boundaries and mingling and working together. As this mingling becomes accepted and customary, a new cosmopolitan way of life will emerge. People may decide to settle down for their retirement in another country, or they may purchase a second house so they can spend their vacations abroad. The movement goes both ways, of course. In Tokyo and Japan's other major cities even today, the number of workers from abroad is rapidly climbing, and their populations are rapidly being internationalized.

Through that might be called a process of information capitalization, corporations are being multinationalized as never before. It is not surprising, then, that their identities gradually grow less distinct. The building momentum of decodage, meanwhile, causes their own codes to give way. It is the need to cope with this situation that is behind the strong desire within corporations to reestablish their identities.

Japanese companies today are all subject, to greater or lesser degrees, to the workings of an economy now at the bidding of electronics technology, and this has led to the undeniable erosion of their core images. Fighting back against this trend, they are working to grasp and identify firm images. The latent energy of codification, in other words, is gaining momentum, and as a result companies have begun to pursue CI strategies.

One often hears it said these days that an effective CI strategy considerably improves company performance in the market and brings enormous profits. This may be so. But it may not be so. If CI strategy does indeed have that effect, it is because the desire for identity on the part of individuals, which has been undermined by high technology, is now growing stronger. Better company performance through CI strategy is a result of this social phenomenon.

Hiroshi Kashiwagi is a critic specializing in the field of corporate identity.

現代日本のCI—ハイテク社会のCI

柏木　博　東京造形大学助教授

たとえば、イタリアのオリベッティ社や、西ドイツのブラウン社、そしてアメリカのIBMやウエスティングハウス社のCIが、トータル・デザインのための事例として日本のデザイン専門誌などで紹介されるようになったのは、1960年代のことである。とは言え、日本の企業やその他の組織体が、それ以前にはCIにまったく無関心だったかと言えば、もちろんそうではない。それ以前から、シンボルマークやロゴタイプが、企業イメージにかかわっていることはよく知られていた。ちなみに、第二次大戦前の広告とグラフィックデザインの専門誌として、最もポピュラーであった『広告界』(誠文堂新光社刊)の、1936年10月号の特集は、「オリンピックカット集」となっており、オリンピックの各競技の図案化を行っている。このトータルなデザイン計画は、幻となった東京オリンピックのためのものである。したがって、日本でCIということがいつ頃から始まったのかを確定することはなかなか困難である。

しかし、コーポレート・アイデンティティ(CI)というコトバをともなって、日本の企業やその他の組織体、そしてデザイナーに、まさにCIということが広がったのは1970年代になってからのことではなかろうか。つまり、それ以前からCI的な戦略を持つデザイン活動はあったわけだが、正しくCIというコトバを使ったデザイン活動が広がったのは、この時代あたりからのことであろう。CIというコトバが定着するにしたがって、企業がシンボルマークやロゴタイプなどをはじめとして総合的なイメージの統一化を戦略的に行うことが目立って増え始めた。

そして、1980年代に入った頃から、過剰だと思えるほどにCIということがさかんに語られ、企業や組織体が、それまでのシンボルマークやロゴタイプを変更したり、時としては建築物をはじめとした空間イメージまで新しいものにして統一化を計ろうとしている。これほど過剰にCIが語られ、また実践されたことはかつてなかったと言っていい。

CIの過剰化とテクノロジーの変化

ではなぜ、80年代に入って以降、日本ではこれほどCIが語られ、また実践されているのか。直接的には、市場における企業イメージ、また市民に対する公共団体のイメージを、ますます強固に印象づける必要性に迫られて来たからだと言えよう。そして、そうした必要性とは、企業や公共団体の統一イメージを強固に主張しない限り、曖昧でぼやけたものになってしまうほど、様々なイメージと情報が氾濫しているという今日の状況の中から出て来たのだととりあえずは言える。また、これもひとつの直接的要因だと言えるが、80年代に入ったあたりから、なんとはなしに、時代の感覚が急速に変化し、それにつれて、企業や公共団体のイメージを新しく変化させる必要性に迫まられたために、CIが過剰に語られ実践されたと言えよう。しかしなぜ、80年代に入ったあたりから、時代の感覚が急速に変化したのか。さらに言えば、本当に時代の感覚は急速かつ大きく変化したのか。こうした疑問に対しては、時代の感覚は急速かつ大きく変化したと即答することができる。つまり、ひとことで言えば、かつてのマシン・テクノロジーから電子テクノロジーの時代へと変化したのである。そして、それにつれて時代の感覚が急速に変化したのだと言えよう。

様々なイメージと情報が氾濫し、CIを強固に表現する必要に迫られたり、また、新しい時代の感覚にあわせてCIを戦略的に設定しなおしたりする必要に迫られていることの、基本的な要因として、実は、マシン・テクノロジーから電子テクノロジー時代への技術の転換があったのだとすら言える。

わたしたちの社会にあっては、技術は、ただ単に、冷たく無表情なものとしてあるのではない。わたしたちの社会は、その隅ずみに至るまで徹底して市場経済(資本)の論理によって支配されている。そして、資本の論理もまた、ただ単に経済活動の論理としてあるのではない。それは、文化の支配的な論理でもある。そして、そうした市場経済の論理は、つねに、テクノロジーの進展と密接に結び付いて来た。したがって、わたしたちの社会にあっては、テクノロジーは、市場経済のあり方とともに文化のあり方と相互にかかわっているのである。

技術と市場経済と文化とのそうした相互の関係が、時代の感覚、そしてそれを具体化したデザインを決定していることの実例を、わたしたちはモダンデザインの歴史の中にいくらでも見ることができる。たとえば、両大戦間の時代は、機械テクノロジーが日常生活の中へと入り込んで行った時代である。機械テクノロジーが社会の中に入り込んで行ったのは、19世紀の産業革命からのことだから、両大戦間の時代よりもずっと早い。しかし、わたしたちの日常生活の中に、様々な道具として入り込んで来たのは両大戦間の時代であった。機械的テクノロジーは、様々な要素を機能主義的に捉え、その要素間の関係をできるだけ合理化しようとする。その合理化とは、言い換えれば"経済化"ということである。したがって、それはまさに経済の論理と見合っていた。そして、機械テクノロジーは、ものを経済的かつ大量に生産しようとした。近代社会の一つの特徴は"大衆"という存在が産み出されたところにある。それは、経済的かつ大量にものを生産し、市場に供給する機械テクノロジーのシステムを背景に産まれて来たと言えるだろう。そうした、機械テクノロジーの論理はまた、グラフィックから建築空間にいたるまで、機械的合理主義(経済主義)にもとづいた、機能的なデザインによって、自らの姿を現わしたのである。

ひとことで言えば、機械テクノロジーは、経済性ということとかかわっていた。と同時に、大量生産・大量消費をめざしつつ、大衆という存在とかかわっていた。そして、それは、機械的論理を反映した、画一的なデザインを生み出したのである。

戦後、いわゆる高度経済成長期に、日本は大衆消費社会化をむかえる。そして、その背景にあったのは、言うまでもなく、機械テクノロジーであった。その機械テクノロジーがいま、電子テクノロジーに変わろうとしている。機械テクノロジーは、いまや過去の、つまりデッド・テック化しようとしているのだ。しかしもちろん、機械そのものがまったく姿を消してしまうということではない。要するに支配的なテクノロジーではなくなったということである。過去の支配的テクノロジーがデッド・テック化し始めたのは80年代に入ったあたりからのことである。それは、日本においてCIが過剰なまでに注目され始めた時期と重なっている。そして、その二つはけっして相互に無縁ではなく、強くかかわりあっているのである。

希薄化するアイデンティティ

2年ほど前に、外国人ジャーナリストから日本のデザインについていくつか質問を受けたことがある。その質問の中の一つは、80年代に入って以降の日本のデザインの特徴をきわめてよく捉えたものであった。その質問はこういうものであった。たとえば、西ドイツのブラウン社やオランダのフィリップス社の家電製品のデザインを例にとると、音響製品からシェーバーに至るまで、どの製品をとって見ても、それぞれ統一感があり、一目でブラウンのものであるとか、フィリップスのものだとわかるが、日本の家電メーカーの製品は、各製品間の統一感はないし、それぞれの製品はメーカーのイメージを特徴づけていない。そして、そういうデザイン傾向は、このところますます強くなって来ていると言うのだ。たしかに、一つのメーカーがつくっている、オーブン・トースターを見ても、ピンクやエメラルドグリーンのポップなものから、無機的でハイテク(ハイスタイル・アンド・テクノロジーの略としてのハイテク)なイメージのものまで、まったく統一感はない。したがって、製品デザインが企業の統一したイメージをつくっているということもない。そして、その傾向は確かに強くなっている。

製品デザインのこうした多様化は、この間、マーケティングの分野で「少量多品種」型の生産と言われたような生産方式によって出て来た現象である。この多様化は、それ以前からある製品デザインの不統一さとは論理的にはまったく別の問題である。すでにふれたように、機械テクノロジーを背景にして、わたしたちの社会は大量生産・大量消費の形態をつくり出した。そうした中で産み出されるものは画一化されたデザインであった。数百万着の単位でつくり出される大量生産品のTシャツはその代表的な例になろう。そうしたものが市場に出てくることによって、誰もが同じTシッツを着て、同じような格好になる。しかし、やがてわたしたち大衆は、誰もが同じであるという画一化され均質化された状態から逸脱したいという欲望を強く持つようになって来た。その欲望が顕在化したのは、少量多品種型の、かつてのものよりも圧倒的に差異の多い商品がマーケットに出始めたことによっている。つまり、潜在的には差異を欲望していたわたしたちの気分を、少量多品種型の商品が顕在化し、また追い打ちをかけたと言えるだろう。この少量多品種型の商品の最も象徴的なものは、すでに経済学者の岩井克人が『ヴェニスの商人の資本論』(筑摩書房)の中で議論しているように、アメリカでつくられた「キャベツ人形」であろう。つまり、少量多品種型の生産を可能にしているのは、コンピューターを始めとする電子テクノロジーなのだ。はたして、今日の少量多品種型の生産は、今後、さらに徹底するかどうかは確定できないが、電子テクノロジーを背景にして、エレメント(部品やパーツ)の数を増し、その組み合せのバリエーションを増して行くことは、ますます進むだろう。

電子テクノロジーがわたしたちの日常生活の中に入り込むことによって、たとえば、少量多品種型の、様々な消費者イメージにあわせたデザインが出現してくる。すると、その製品をつくり出している企業は、多様化する製品デザインによって、もはや統一した企業イメージを表現することはきわめて困難にならざるをえない。そうした事態の中にあって、企業のイメージは、希薄化する。

あらためて言うまでもないが、デザインつまりde-signとは記号にかかわる作業である。そして、今日、デザインはもはや記号性を通り越して“情報”と化しているとすら言えよう。情報であるからには、新しさとスピードが要求される。その結果、次からつぎへと新しいデザインが、その多様性を主張して出現してくる。つまり、情報(デザイン)が氾濫する中にあって、今日、企業や公共団体は、統一イメージを強固に打ち出さないかぎり、自らのイメージが希薄なものになってしまうのである。その結果、統一感のある“象徴”的なイメージをCI戦略によって打ち出そうとする欲求が強くなって来ているのだと言えるだろう。

ところで、経済の動きは、その純粋な論理にしたがって行けば、当然のことながら、無限に自己を拡張しようとする方向へとむかって行く。だから論理的には、経済の動きは、自らにとって障害となるものはすべてはねのけようとする。たとえば、宗教的障害や道徳的障害、あるいは社会的な通念といった障害をすべてはねのけ、自らを拡大、拡張しようとする。人間がほっておいても、自律的に動いて行くような力を持っている。あらゆる規範をはねのけて、自己を拡大、拡張しようとする経済の動きを、“脱コード”化の運動と捉えることもできる。

とは言え、わたしたちの社会は、そうした経済の動きを放っておくのではなく、つねに新たに再編しなおし、一定の規範の中に囲いこみ、再びコード化しようとしてきた。つまり、経済の動きは、いつもコードを逸脱しようとする力と、それを再びコード化しようとする力とのせめぎあいの中にあったと言えるかもしれない。そして、それはつねに変化して来たと言えるだろう。

今日、そうした経済の動きは電子テクノロジーを背景にして、あきらかに国境という囲いを突破しつつある。してみると、資本には国境も国籍もなくなってしまうのだろうか。おそらく、論理的にはそうなのだろう。その良し悪しは別にして、今日の多国籍企業の姿や、世界を同時に結ぶ情報のネットワークの拡張を見ればそのことはすぐにわかる。資本が今後、徹底して国境も国籍も失うほどになってしまうとはもちろん思えない。その前になんとか、規範の中に収めようとする力が働くだろう。

しかし、少なくとも、日本はいまハイテク化され、企業は国境を越えて国際化されつつある。資本や情報が国境を越えるということはまた当然のことながら、人間も国境を越えて入り乱れて仕事をすることになる。そうした現象が日常化するにつれ、それが成功するか否かは別として、老後は外国で過そうとか、外国にセカンドハウスを持とうとかといった、日常感覚が出てくる。また、逆に、現在の東京のように、外国からの労働力が入り込んで来て、無国籍的な性格を少しづつ強めているということもある。

いわば情報資本主義化によって、企業は、かつてとは比較にならないほどにまで、多国籍化しつつある。その結果、自らのアイデンティティが必然的に希薄化するといったことが出てくる。脱コード化すればするほど、コードが希薄化するのだから当然である。そうした、今日の状況の中にあっては逆に、自らのアイデンティティを強く望む気分が出て来ているとも言えるだろう。

つまり、大なり小なり、今日の日本の企業は、電子テクノロジーが支配する経済の動きの中にあって、自己の中心的なイメージが希薄化して行っていることは否定しえない。そうした中で、なんとか自己の中心的な、確固としたイメージを掴んでおこうとする、つまり再びコード化しようとする力が潜在的に働き始める。その結果として、CI戦略的なものを求めていると言えないだろうか。

CIをやるとマーケットの効果が上り、多くの利潤につながるといった言い方を最近よく耳にする。それは本当かもしれない。あるいはそうではないかもしれない。しかし、もしそうだとすれば、それは、わたしたちの社会がハイテクの支配によって、アイデンティティを希薄化させており、だからこそまた、それを求めようとする人々の欲望が強くなっているからなのではないか。したがってCIが利潤につながるか否かは、そうした社会的な現象の一つの結果でしかあるまい。

WORKS

1a

1a~1f
NTT CI Plan
NTTCI計画

Consulting
PAOS Inc.
株式会社パオス
Motoo Nakanisi
中西元男
Dentsu Inc.
株式会社電通
Designer
Yusaku Kamekura
亀倉雄策
Client
Nippon Telegraph
and Telephone Corporation
日本電信電話株式会社
1985

Designing a corporate identity mark for NTT, Japan's biggest corporation, would make even the most audacious person shrink. But I was determined to take up the challenge. "Our mark should be simple, impressive, yet have a space-age image suggesting optical fibers," said the NTT president.
(Yusaku Kamekura)

The privatization of the government-run communications monopoly after a history of 115 years gave birth to Japan's largest private corporation, Nippon Telegraph and Telephone, perhaps the greatest event in the history of Japanese enterprise. The advent of this new corporate identity, moreover, presented one of the most momentous projects in the history of the CI planning profession in Japan.
(Motoo Nakanishi)

日本最大のマンモス会社 NTT のマークをデザインすることは、どんな神経の太い人でも萎縮してしまうに違いない。巨大なスケールをどう具象化するか、それだけの格調を創り出せるか。その重圧をはね返さないかぎり、私の勝負はない。真藤総裁は、「マークはシンプルで印象的で大衆に親しめ、しかも光ファイバーなどを含んだ宇宙的なイメージが欲しい」と注文された。言葉では簡単だが、視覚化には不可能といっていい注文だった。
(亀倉雄策)

NTTのCIは、115年の長きにわたり官営であった電気通信事業が民営化され、日本最大の私企業が誕生するという、わが国企業史上の大事件であった。そればかりでなく、結果からみると日本のCI史を飾る大プロジェクトでもあった。NTTのCIを通じて、いかに多くの人達がCIを知り、関心を抱き、その成果に注目したか、今さら論をまたない。これを支えたものが亀倉雄策氏の優れたマークデザインと、そのシステマティックなデザイン展開ではなかったかと考えている。
(中西元男)

NTT

1b

1c

1d

1e

1f

2a

2a～2d
RECRUIT CI Plan
リクルートCI計画

Designer
Yusaku Kamekura
亀倉雄策

Client
Recruit Co., Ltd.
株式会社リクルート
1987

2b

Recruit adopted the seagull mark in its eager-beaver days as a new company, but today it is a mammoth enterprise. Now a leader in the fast-growing information industry, it has departments handling remote computing service and information network system. The company is seeking a suitable visual expression for its new thrust in operations, and has just embarked on a project to determine how the seagull mark can be adapted to a new corporate image.
(Yusaku Kamekura)

マンモス会社に成長したリクルートは、若さいっぱいの時代を象徴したカモメだが、今では巨大な情報企業という内容に変貌しつつある。既にコンピューター事業（リモート・コンピューティング・サービス事業）や情報ネットワークサービス事業といった新しい形態がそれぞれに視覚表現を求めており、それらの情報表現をいかにしてカモメと組み合わせるかという作業が、今始まったばかりである。
（亀倉雄策）

2c

2d

3a～3h
Surbank Setagaya CI Plan
サバンク世田谷CI計画

Consulting
PAOS Inc.
株式会社パオス
Designer
Yusaku Kamekura
亀倉雄策
PAOS Inc.
株式会社パオス
Client
The Setagaya Shinkin Bank
世田谷信用金庫
1986

The Setagaya Credit Union's image was so tarnish and outdated that it had ceased to draw new customers from the younger generations. It would have to make a complete about-face in order to survive, its management realized, and they hired graphic designers Yusaku Kamekura and Motoo Nakanishi to do the job. Thanks to Kamekura's tulip logo and the total consulting system devised by Paos, Inc., Surbank Setagaya not only got a new lease on life, it showed top performance among financial concerns of its kind in Tokyo.
(Yusaku Kamekura)

In marking the 65th anniversary of its founding, the Setagaya Shinyo Kinko (Setagaya Credit Union) introduced a new corporate image, changing its advertising name from *Seta Shin* to *Surbank Setagaya*. Its corporate character is symbolized by the tulip mark. The major corporate color is green, red, gray and white are also used. The rounded contours of the logo are chosen to express the friendliness, warmth, and up-to-date character of this suburban banking concern.
(Motoo Nakanishi)

3a

世田谷信用金庫というと、もう時代遅れで若い人を吸収する力を次第に失ってきていた。そこで、まさに180度の大旋回をしなければ活路がないという決心のもとに、亀倉と中西元男のコンビで作業を始めた。亀倉のチューリップのマークのロゴや展開をパオスが完全なシステムで実行し、この信用金庫を蘇らせるどころか見事に東京一の成績にすることに成功した。
(亀倉雄策)

3b

創立65周年を機に導入されたCIシステム。せたしんからサバンク世田谷へとコミュニケーションブランドの変更を行うとともに、コーポレートキャラクターにチューリップをシンボライズ。またコーポレートカラーは、グリーン及びレッド、グレー、ホワイトを併用するマルチ展開方式を採用。ロゴタイプも丸味を生かした造形で、郊外型ニューバンクビジネスとしての親しみやすさ、暖かさ、進んだ感覚の表現を図っている。
(中西元男)

3c

3d

3e

3f

3g

3h

4a〜4e
ICSID '73 Kyoto Symbol Mark
ICSID '73 Kyoto シンボルマーク

Designer
Yusaku Kamekura
亀倉雄策
Client
International Council of Societies of Industrial Design.
世界インダストリアルデザイン会議
1973

4a

4b

4c

ICSID is an international group of industrial designers. This mark was created for its convention when it met for the first time in Japan (Kyoto). Like industrial design, the mark is three-dimensional, a volume built out of the ICSID '73 acronym and set within a series of concentric rings. The number of rings increases in three stages leading up to the opening of the meeting. The logo with the largest number of rings represents the point at which the convention actually began.
(Yusaku Kamekura)

ICSIDは国際的なIDデザイナーの団体である。この団体の会議が我が国で初めて京都で行われ、大会用のシンボルマークが作られた。IDを表す視覚の立体をICSIDという文字で構成し、それに広がりの輪を三段階に増やしていくというアイデアである。一番輪が増えた時点が、ちょうど大会開催の時期になる。
(亀倉雄策)

4d

4e

5a〜5b

The Association for the Promotion of Traditional Craft Products Symbol Mark

伝統的工芸品産業振興協会 シンボルマーク

Designer
Yusaku Kamekura
亀倉雄策

Client
The Association for the Promotion of Traditional Craft Products.
財団法人伝統的工芸品産業振興協会
1975

5a

5b

The Association for the Promotion of Traditional Craft Products is an organization devoted to active efforts to encourage the use of traditional craft products in daily life. The top Half of the mark is made up of a stylized figure based on the first letter "den" of the word dentō (tradition). The other half contains a red "sun" symbolic of Japan. A feature of this mark is that it can be used in any direction, with the two parts freely arranged in a vertical or horizontal design.
(Yusaku Kamekura)

伝統的工芸品産業振興とは、日本の手工芸の美しさを日本人の生活に積極的に浸透させせようという運動である。そこで伝統の伝の字を形象化して日本を象徴する赤い丸を配した。このマークは縦でも左横でも右横でも自由に使用できるのが特徴である。
(亀倉雄策)

6a〜6b

Morisawa Logotype

モリサワ ロゴタイプ

Designer
Yusaku Kamekura
亀倉雄策

Client
Morisawa Phototypesetting Machine Co., Ltd.
株式会社モリサワ
1982

モリサワ

6a

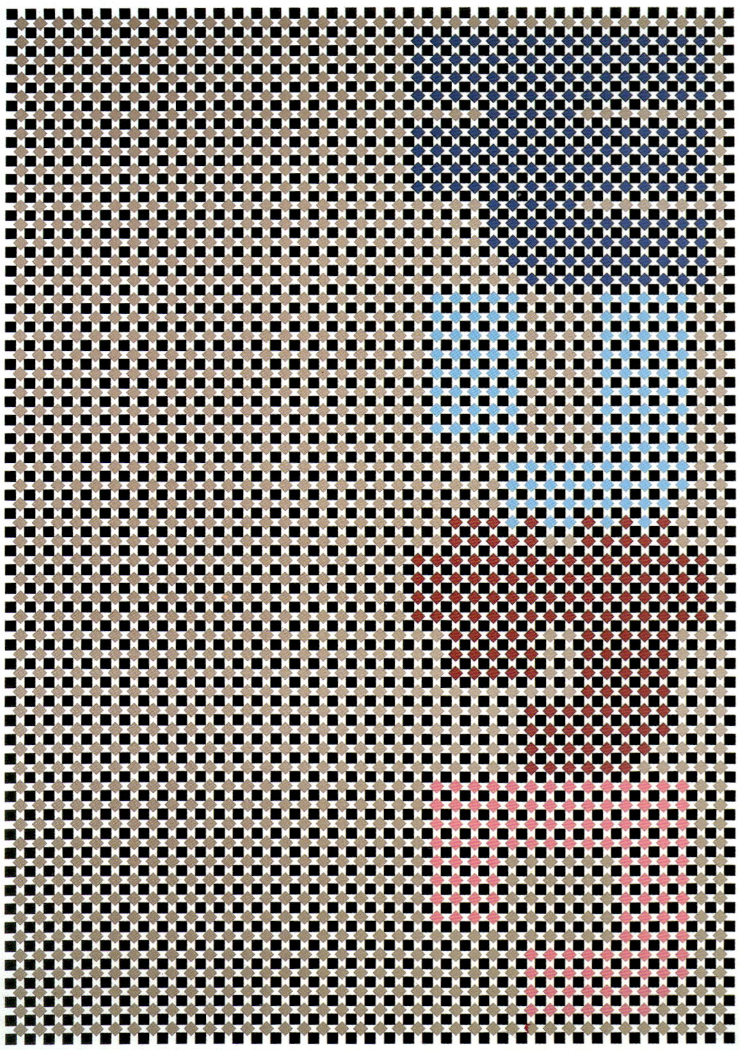

6b

The letter Morisawa is difficult to design in logo partly because it is so very ordinary. This design succeeds in creating a logo that is readable, memorable as well as elegant.
(Yusaku Kamekura)

モリサワという文字をロゴタイプにするのは難しい。あまりにも平凡だからである。可読性があり、見る人にどこかで印象を与えるという特徴を、嫌味なく作ることに成功したロゴタイプである。
(亀倉雄策)

7

8

9a

9b

7
Ark System Symbol Mark
アークシステム シンボルマーク

Designer
Yusaku Kamekura
亀倉雄策

Client
Ark System Co., Ltd.
株式会社アークシステム
1986

8
Takara Shuzo Symbol Mark
宝酒造 シンボルマーク

Designer
Yusaku Kamekura
亀倉雄策

Client
Takara Shuzo Co., Ltd.
宝酒造株式会社
1985

9a〜9b
Nikon Auto-Focus Symbol Mark
ニコンオートフォーカス シンボルマーク

Designer
Yusaku Kamekura
亀倉雄策

Client
Nippon Kogaku K.K.
日本光学工業株式会社
1982

10

11

12

10
Speech Quality Standard for Telephone Set Symbol Mark
電話機通話品質標準規格 シンボルマーク

Designer
Yusaku Kamekura
亀倉雄策

Client
Communications Industry Association of Japan
通信機械工業会
1986

11
Nikon Eye Catch Mark
ニコン アイキャッチマーク

Designer
Yusaku Kamekura
亀倉雄策

Client
Nippon Kogaku K.K.
日本光学工業株式会社
1982

12
World Trade Center Bldg. Symbol Mark
世界貿易センタービル シンボルマーク

Designer
Yusaku Kamekura
亀倉雄策

Client
World Trade Center Building Corporation
株式会社世界貿易センタービルディング
1973

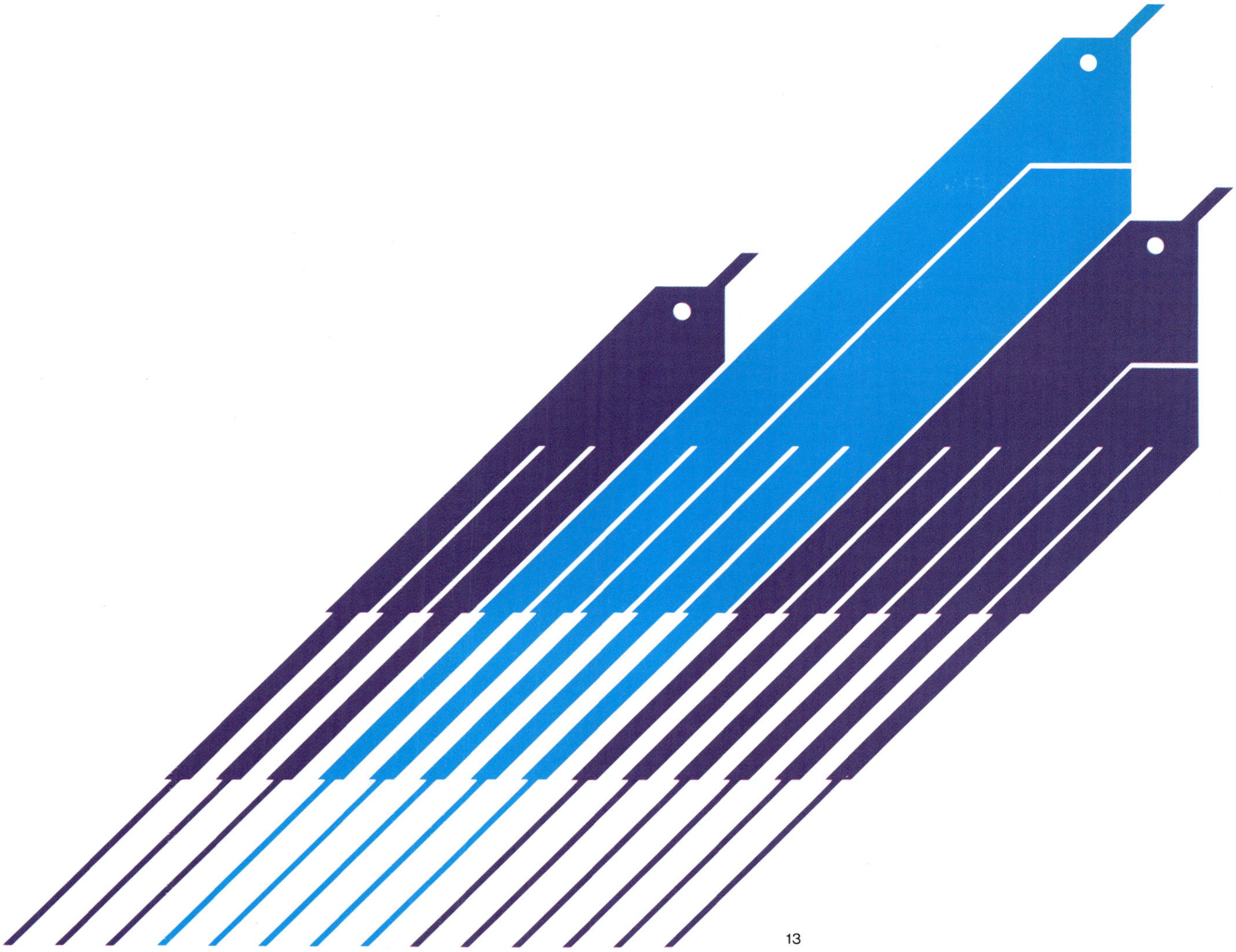

13

14

13
Kanjuku Symbol Mark
關塾 シンボルマーク

Designer
Yusaku Kamekura
亀倉雄策

Client
Kanjuku
株式会社 關塾
1986

14
Kitaura Golf Club Symbol Mark
北浦ゴルフ倶楽部 シンボルマーク

Designer
Yusaku Kamekura
亀倉雄策

Client
Kitaura Golf Club
北浦ゴルフ倶楽部
1986

15

16

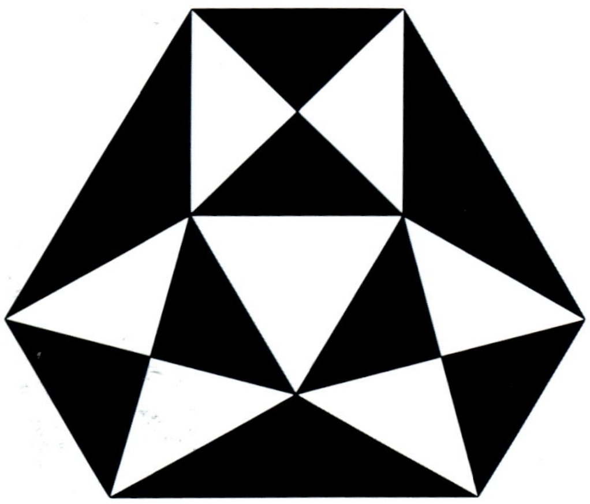

17

18

15
Yamagiwa Symbol Mark
ヤマギワ シンボルマーク
Art Director & Designer
Yusaku Kamekura
亀倉雄策
Client
Yamagiwa Corporation
ヤマギワ株式会社
1966

16
Design Year Symbol Mark
デザインイヤー シンボルマーク
Art Director & Designer
Yusaku Kamekura
亀倉雄策
Client
Japan Industrial Design Promotion Organization
財団法人日本デザイン振興会
1973

17
TDK Symbol Mark
TDK シンボルマーク
Art Director & Designer
Yusaku Kamekura
亀倉雄策
Client
TDK Corporation
TDK株式会社
1966

18
Prince Hotel Symbol Mark
プリンスホテル シンボルマーク
Art Director & Designer
Yusaku Kamekura
亀倉雄策
Client
Prince Hotels Inc.
株式会社プリンスホテル
1968

19 20 21

22 23 24

25 26 27

19
Toho Shigyo & Toho Kako Symbol Mark
東邦紙業 東邦加工 シンボルマーク
Art Director & Designer
Yusaku Kamekura
亀倉雄策
Client
Toho Shigyo Co., Ltd.
Toho Kako Co., Ltd.
東邦紙業株式会社, 東邦加工株式会社
1964

20
Zenkyoren Symbol Mark
全共連 シンボルマーク
Art Director & Designer
Yusaku Kamekura
亀倉雄策
Client
National Mutual Insurance Federation of Agricultutal Co-operatives
全国共済農業共同組合連合会
1973

21
Fuji Bank Symbol Mark
富士銀行 シンボルマーク
Art Director & Designer
Yusaku Kamekura
亀倉雄策
Client
The Fuji Bank, Ltd.
富士銀行株式会社
1979

22
Tostem Logotype
トステム ロゴタイプ
Art Director & Designer
Yusaku Kamekura
亀倉雄策
Client
Toyo Sash Co., Ltd.
トーヨーサッシ株式会社
1985

23
The Ginza Centennial Aniversaly Symbol Mark
銀座100年記念 シンボルマーク
Art Director & Designer
Yusaku Kamekura
亀倉雄策
Client
Ginza Merchants Association
銀座通連合会
1967

24
11th Interski Zao, Japan '79 Symbol Mark
第11回インタースキー蔵王大会 シンボルマーク
Art Director & Designer
Yusaku Kamekura
亀倉雄策
Client
The Organizing Committee of The 11th interski Zao Japan
第11回インタースキー蔵王大会実行委員会
1977

25
Global Eye Design Exhibition Symbol Mark
グローバルアイ展 シンボルマーク
Art Director & Designer
Yusaku Kamekura
亀倉雄策
Client
Japan Design Committee
日本デザインコミッティ
1969

26
Japan Architects Association Symbol Mark
日本建築家協会 シンボルマーク
Art Director & Designer
Yusaku Kamekura
亀倉雄策
Client
Japan Architects Association
社団法人日本建築家協会
1960

27
Kyowa Hakko Symbol Mark
協和醱酵 シンボルマーク
Art Director & Designer
Yusaku Kamekura
亀倉雄策
Client
Kyowa Hakko Kogyo Co., Ltd.
協和醱酵工業株式会社
1967

28a～28b
Tsukuba EXPO '85 CI Plan
科学万博ーつくば'85CI計画

Art Director & Designer
Ikko Tanaka
田中一光
Illustration – 28b
Tetsu Uehara
上原　徹
Client
Japan Association for The International Exposition, Tsukuba, 1985
財団法人国際科学技術博覧会協会
1985

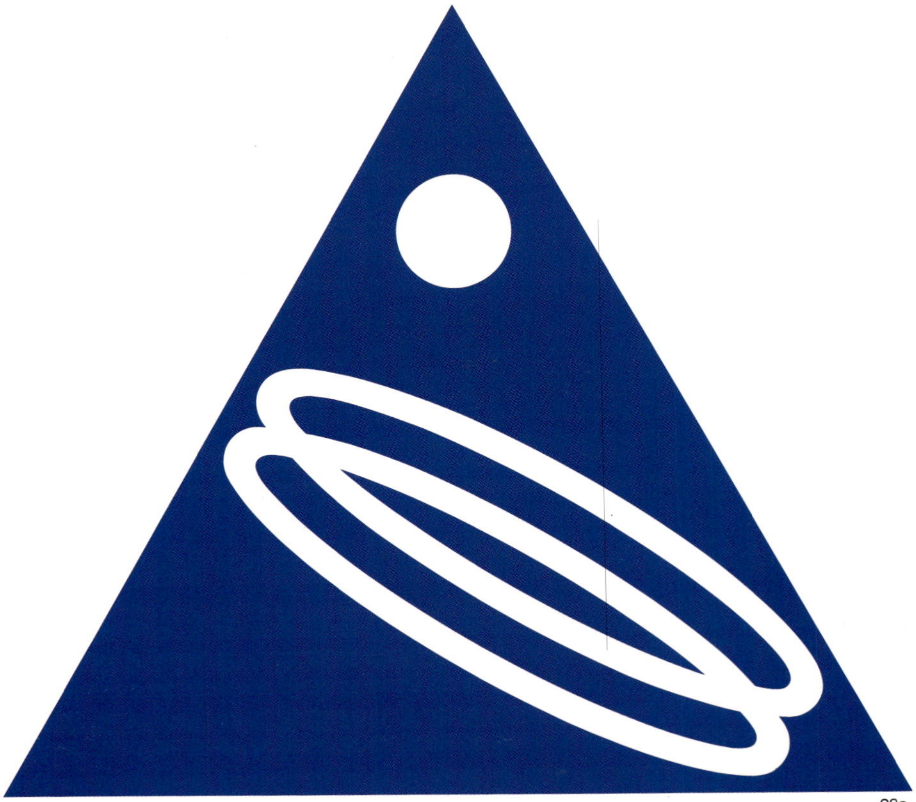

28a

Symbol and corporate marks are being produced in ever-increasing numbers these days, and this design is based on an original sketch. The triangle, I thought, best symbolized the intellectual and neutral character of science. Its pinnacle stands for the peak of Mt. Tsukuba in whose shadow Expo '85 was held. The three sides of the triangle symbolize the themes of the exposition: man, habitat, environment. The parallel rings express the harmonious orbiting of science and human affairs. (Ikko Tanaka)

マークが氾濫している現在、ありそうなマークは除外し、白紙の状態から絵をかくつもりでデザインした。特に科学のもつ知的クールさは、三角形が最もふさわしいと思った。三角形の頂点は開催地である筑波山の峰を象徴し、三辺は博覧会のテーマである人間、居住、環境の断面を、平行する二つのリングは人間と科学との調和した軌道を表す、ということにした。
（田中一光）

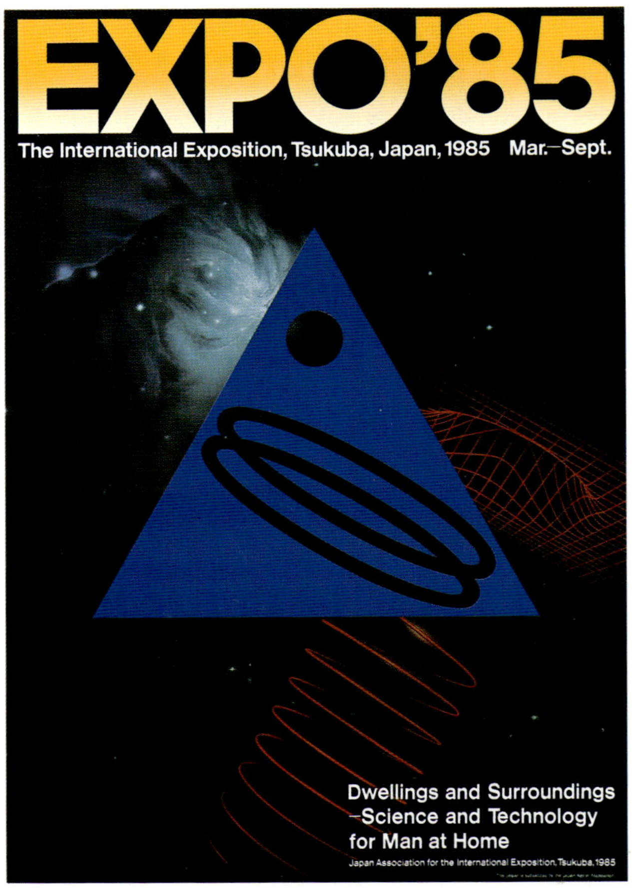

28b

29a~29b
Silk Road Exposition Nara '88. VI Plan
なら・シルクロード博VI計画

Art Director
Ikko Tanaka
田中一光

Designer
Masaaki Hiromura
広村正彰

Illustration – 29b
Akira Mori
毛利 彰

Client
Association for Silk Road Exposition, Nara '88.
財団法人なら・シルクロード博協会
1985

29a

The Silk Road formed a major link between the civilizations of East and West from ancient times. The curvilinear motif evolved in many design forms in Japan. One is the arabesque pattern familiar among Japanese as a common pattern on furoshiki wrapping cloths. This pattern, closely resembling both the figure 8 and the letter S, seemed fitting to an exposition to be held in the year 1988 and to the theme: 'Silk Road.' The shape, with scroll at top and bottom, represents the dynamic and strong tie between East and West.
(Ikko Tanaka)

シルクロードは、東洋と西洋の文明を結ぶ太いきずなである。当時からの曲線模様は、我が国でも様々な意匠に展開された。中でも我々になじみ深いのは風呂敷の唐草模様である。この唐草をモチーフにして1988年に開催される8の字とSilkのSの字にも見えるよう、また上下の渦巻状のかたちは、東と西を結ぶ太いきずなを表現したものである。
(田中一光)

29b

30a〜30c
Iwate Broadcasting CI Plan
岩手放送CI計画

Art Director & Designer
Ikko Tanaka
田中一光

Client
Iwate Broadcasting Co., Ltd.
岩手放送株式会社
1984

30a

The mark of Iwate Broadcasting Co., Ltd. IBC is based on the letter 'I' taken from the corporate name of IWATE. The oblique oval is intended to suggest of the parabola antenna, the orbit of satellite, and the solar system. The standing rectangle symbolizes the steadfast management of IBC. The mark appears well balanced with the 'movement' of the oval and the 'stillness' of the rectangle. IBC has great enthusiasm concerning its CI and we were very successful because of close communications. I keenly felt the importance of mutual eagerness between the creator and the firm itself in successful CI. (Ikko Tanaka)

このマークは、IWATEのIを形象化した。傾斜した動きのある楕円形は、パラボラアンテナ、人工衛星、さらには太陽系をも暗示させるように考えた。直立した矩形は、岩手放送という安定した企業の象徴で、楕円の〈動〉と矩形の〈静〉の均衡によって成り立っている。岩手放送では、たいへん熱心にCIに取り組まれ、密度の高いコミュニケーションが図られた。CIには、つくり手と使い手の連帯感が大切であると痛感した。
(田中一光)

30b

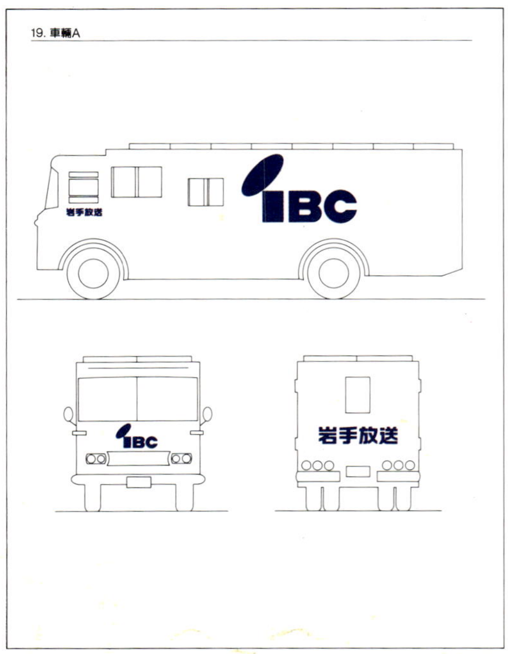

30c

31a〜31c
Issey Miyake licence group BI Plan
イッセイ・ミヤケ・ライセンスグループ BI計画

Art Director & Designer
Ikko Tanaka
田中一光

Aprecation
Break Fast
ブレック・ファースト

Client
Miyake Design Studio.
株式会社三宅デザイン事務所
1979

This logo, made up of the initials i and m, was designed as a symbol for the Issey Miyake line of licensed products. At first I thought it might have been too rigid a design for use as a symbol for fashion merchandise and boutiques, but in retrospect it seems to be quite fitting. The applications were all created by Break Fast.
(Ikko Tanaka)

i と m は、Issey Miyake のイニシャルで、三宅一生のライセンス・プロダクトのシンボルである。ファッション商品やブティックのシンボルとして多少硬質な感じがするのではと思っていたが、今となってはこれでよかったような気がする。アプリケーションはすべて〈ブレック・ファースト〉の手によって行われている。
(田中一光)

31a

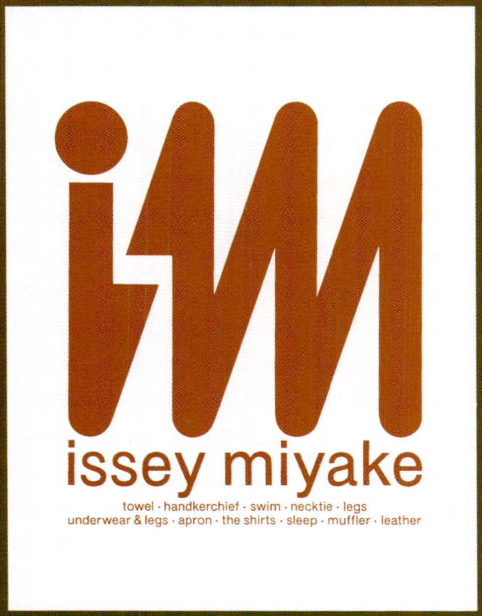

31b

31c

32a〜32e
Seibu Saison Group CI Plan
西武セゾングループ各社CI計画

Creative Director
Ikko Tanaka
田中一光
Client
Seibu Saison Group
西武セゾングループ
1982〜1986

SAISON

32a

32b

The Seibu, Group, like so many corporations which have been actively diversifying their operations since the beginning of the 1980s, has long since ceased to be simply a "distribution" enterprise. Because there was no knowing what kind of new business it might extend itself into or in what direction its enterprises might develop, it needed to adopt a unifying logo. Initially used for a credit card, the SAISON logo was eventually elevated to the status of logo for the entire complex of enterprises as the Group began to spread out in the culture industry. The logo, Futura typeface lettering arranged in an original design, was planned to serve as the official mark for the Seibu Group.
(Ikko Tanaka)

1980年代に入ると、企業活動の多角化がいっそう活発になって、西武グループの体質もすでに流通という言葉に収まり切らなくなった。どんな業種が生まれ、何が起こるか解らないこのグループのフレキシビリティを考えると、統一書体の制定が必然的に起った。SAISONというロゴも、当初、クレジット・カードのためにデザインしたものだが、文化事業を中心に拡大され始め、いつの間にか、グループのシンボルロゴにまで昇格してしまった。当然、このロゴの基本であるフーツラをアレンジした書体をこのグループの制定文字として計画した。
(田中一光)

銀座セゾン劇場
GINZA SAISON THEATRE

32c

TICKET
チケット・セゾン
SAISON

32d

32e

33a

33a～33d
Mujirushi Ryohin BI Plan
無印良品BI計画

Creative Director
Ikko Tanaka
田中一光

Graphic Designer
Hiroshi Kojitani
麹谷 宏

Interior Designer
Takashi Sugimoto
杉本貴志

Illustrator
Shigeo Fukuda
福田繁雄

Copywriter
Kazuko Koike
小池一子

Client
The Seiyu, Ltd.
株式会社西友
1981～1985

33b

The Seiyu's original line of Mujirushi Ryohin (No-brand Merchandise) is low priced because of a new concept based on careful selection of materials, close inspection in the production process, and simplification of packaging. The consumer's right to know why a product is cheap and how it was made is another important idea behind this line of goods. No-Brand goods, sold without relying on elaborate commercial designs, have become associated with the most up-to-date way of living, and have carned a fashionable image among consumers who prize simplicity and genuineness. (Ikko Tanaka)

西友のオリジナル商品である無印良品は低価格であるための三つのわけとして、素材の選択、工程の点検、包装の簡略化というコンセプトがあって、それらが「わけあって安い」という言葉に集約されていくのだが、なぜ安いのか、またどういう作られ方をしているのかということをあからさまに知らせるということも、この商品の重要な思想となった。商業主義的デザインを一切否定したこの商品群は逆に最も今日的なライフスタイルを代表するひとつとして、ファッショナブルな評価さえうけるようになった。
(田中一光)

33c

33d

34a〜34d
Japan Creative BI Plan
ジャパン・クリエイティブBI計画

Creative Director
Ikko Tanaka
田中一光

Art Director & Designer
Katsuhiro Kinoshita
木下勝弘

Copywriter
Kazuko Koike
小池一子
Momoko Takeuchi
竹内桃子

Photographer –34c
Taishi Hirokawa
広川泰士

Client
The Seibu Department Stores, Ltd.
株式会社西武百貨店
1985

34a

34b

34c

Japan Creative features products that display Japanese originality, satisfy our desire for things that are aesthetically pleasing, and perform a needed function in modern life. This line of goods made its debut at Seibu's Shibuya Department Store in the autumn of 1985 with items by artists whose work goes a step beyond the genres of traditional crafts, art, and design. The black and white symbol mark combining the brush-stroked calligraphy JC and Japan Creative in Gothic type was adapted for use on packaging, and for advertising and signs.
(Ikko Tanaka)

日本のオリジナリティを基本とし、しかも精神的にも満たされ、現代人の生活に十分機能できるような製品がJCのコンセプトである。ジャンルを超えた出品者の手によって、1985年秋、西武渋谷店の全店イベントとしてデビュー。毛筆のカリグラフィとゴチック体をコンバインしたブラック&ホワイトのシンボルマークは、パッケージ、広告、サインなどにも展開された。
(田中一光)

34d

35a～35g
Ciao ITALIA BI Plan
チャオ・イタリアBI計画

Art Director & Designer
Ikko Tanaka
田中一光

Copywriter
Shunichi Iwasaki
岩崎俊一

Client
The Seibu Department Stores, Ltd.
株式会社西武百貨店
1986

Ciao ITALIA was a large-scale exhibition of Italian goods held at Seibu Department Stores throughout Japan in 1986. The symbol mark evokes the sunny land of Italy and the jovial image of the Italian people. The design was used for wrapping materials, shopping bags, special Ciao ITALIA merchandise, and store decorations for the event.
(Ikko Tanaka)

チャオ・イタリアは、1986年全国の西武百貨店で展開された大規模なイタリア展である。このチャオ・イタリアは、太陽の国イタリアを象徴するシンボルを、との要望があり、イタリア的に陽気な太陽がデザインされている。それらは包装、バッグ、関連商品、店内装飾などに展開され、様々なイベントにも及ぶものとなった。
(田中一光)

35a

35b

35c

35d

35e

35f

35g

36a～36b
Hanae Mori CI Plan
ハナエ・モリCI計画

Art Director & Designer
Ikko Tanaka
田中一光
Client
Hanae Mori Tokyo Co.
ハナエ・モリ・トーキョー
1978

36a

This butterfly is the well-known trademark of Hanae Mori. The mark was created for use at the time of the opening in 1978 of the Hanae Mori Building designed by architect Kenzo Tange in Tokyo's Omotesando and for its interior motifs and posters. Mori took a great liking to the butterfly with its elongated, triangular wings, and used it thereafter for her packaging and stationery. It wasn't long until the design had become her own personal CI mark.
(Ikko Tanaka)

いうまでもなく、蝶はハナエ・モリの象徴ともいえる重要なモチーフである。このマークは1978年、東京、表参道に丹下建三の設計によるハナエ・モリビルが竣工する際に、オープニング・シンボルとして、ポスターや店内サインなどのためにデザインしたものである。この三角形の尖った蝶を、森英恵さんが大そう気に入って下さって、その後もパッケージやステーショナリーにも使用され、いつのまにかハナエ・モリのCIマークのようになってしまつた。
(田中一光)

36b

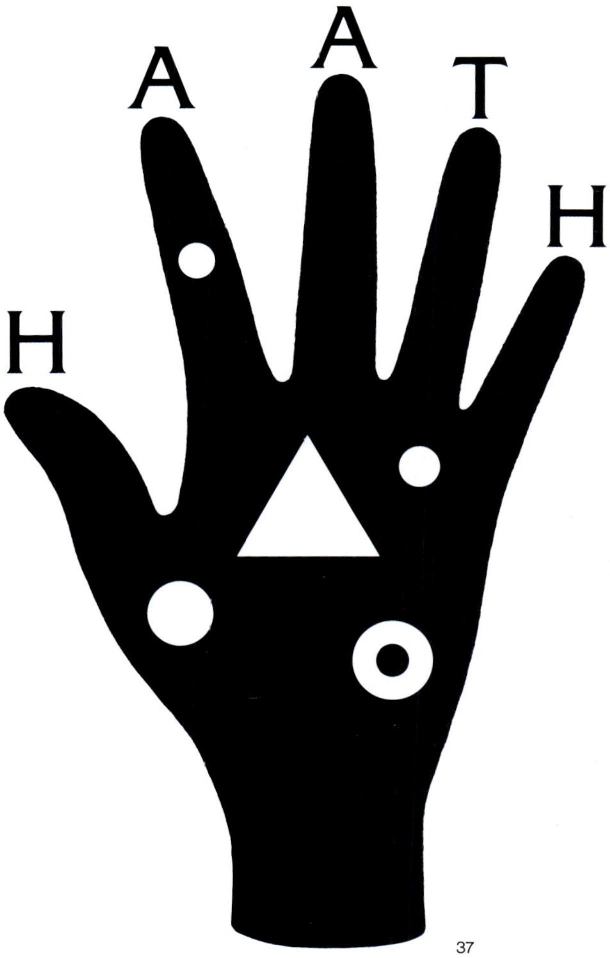

37

38

39

37
HAATH Symbol Mark
ハート・インドの手 シンボルマーク

Art Director & Designer
Ikko Tanaka
田中一光

Client
The Seibu Department Stores, Ltd.
株式会社西武百貨店
1986

38
Saison-no-Eikokuhaku Symbol Mark
セゾンの英国博 シンボルマーク

Art Director & Designer
Ikko Tanaka
田中一光

Client
The Seibu Department Stores, Ltd.
株式会社西武百貨店
1987

39
Cash Point Symbol Mark
キャッシポイント シンボルマーク

Art Director & Designer
Ikko Tanaka
田中一光

Client
The Seibu Credit Co., Ltd.
株式会社西武クレジット
1980

40

41

42

43

44

40
Ginza Graphic Gallery Logotype
銀座グラフィックギャラリー ロゴタイプ
Art Director & Designer
Ikko Tanaka
田中一光
Client
Dai Nippon Printing Co.
大日本印刷株式会社
1986

41
I&S Logotype
I&S ロゴタイプ
Art Director & Designer
Ikko Tanaka
田中一光
Client
I&S Corporation
株式会社 I&S
1986

42
Library Aqua Logotype
ライブラリー・アクア ロゴタイプ
Art Director & Designer
Ikko Tanaka
田中一光
Client
Toto Ltd.
東陶機器株式会社
1987

43
OZ Logotype
西友大泉店 オズ ロゴタイプ
Art Director & Designer
Ikko Tanaka
田中一光
Client
The Seiyu, Ltd.
株式会社西友
1983

44
Morisawa's 60th anniversary Symbol Mark
モリサワ60周年 シンボルマーク
Art Director & Designer
Ikko Tanaka
田中一光
Client
Morisawa photo typesetting Machine Co., Ltd.
株式会社モリサワ
1983

45
Koshino Junko Logotype
コシノ・ジュンコ ロゴタイプ
Art Director & Designer
Ikko Tanaka
田中一光
Client
Koshino Junko Design Office Co., Ltd.
株式会社コシノ・ジュンコ デザインオフィス
1984

46
OLD NEW Logotype
オールドニュー ロゴタイプ
Art Director & Designer
Ikko Tanaka
田中一光
Client
The Restaurant Seibu. Ltd.
株式会社レストラン西武
1983

47
Sagacho Exhibit Space Logotype
佐賀町エキジビットスペース ロゴタイプ
Art Director & Designer
Ikko Tanaka
田中一光
Client
Sagacho Exhibit Space
佐賀町エキジビットスペース
1987

48
SPIRAL Logotype
スパイラル ロゴタイプ
Art Director & Designer
Ikko Tanaka
田中一光
Client
Nichinan Co., Ltd.
日南株式会社
1981

49
Heart Art Symbol Mark
ハート アート シンボルマーク
Art Director & Designer
Ikko Tanaka
田中一光
Client
Fujie Textile
フジエテキスタイル
1973

50

51

52

53

50
Bank of Tokyo Symbol Mark
東京銀行 シンボルマーク
Art Director & Designer
Ikko Tanaka
田中一光
Client
The Bank of Tokyo Co., Ltd.
株式会社東京銀行
1967

51
Asahi Mutual Life Insurance Symbol Mark
朝日生命保険相互会社 シンボルマーク
Art Director & Designer
Ikko Tanaka
田中一光
Client
Asahi Mutual Life Insurance Company.
朝日生命保険相互会社
1972

52
Iwatani Symbol Mark
岩谷産業 シンボルマーク
Art Director & Designer
Ikko Tanaka
田中一光
Client
Iwatani International Corporation
岩谷産業株式会社
1970

53
Ohkura Symbol Mark
大倉 シンボルマーク
Art Director & Designer
Ikko Tanaka
田中一光
Client
Ohkura Pulp & Paper Co., Ltd.
大倉紙パルプ商事株式会社
1971

54

55

56

57

54
Daiko Electric Symbol Mark
大光電機 シンボルマーク
Art Director & Designer
Ikko Tanaka
田中一光
Client
Daiko Electric Co., Ltd.
大光電機株式会社
1982

55
Kyodo Oil Symbol Mark
共同石油 シンボルマーク
Art Director & Designer
Ikko Tanaka
田中一光
Client
Kyodo Oil Co., Ltd.
共同石油株式会社
1966

56
Tanabe Museum Symbol Mark
田辺美術館 シンボルマーク
Art Director & Designer
Ikko Tanaka
田中一光
Client
Tanabe Museum
田辺美術館
1979

57
The World Crafts Council Symbol Mark
世界クラフト会議 シンボルマーク
Art Director & Designer
Ikko Tanaka
田中一光
Designer
Tetsuya Ota
太田徹也
Client
The World Crafts Council
世界クラフト会議
1978

58a～58c
The XIth Olympic Winter Games, Sapporo '72 CI Plan
第11回冬季オリンピック札幌大会CI計画

Art Director & Designer
Kazumasa Nagai
永井一正

Client
The Organizing Committee for The XIth Olympic Winter Games, Sapporo
財団法人札幌オリンピック冬季大会組織委員会
1972

58a

The Winter Olympics were held for the first time in Japan in 1972, a major national event succeeding the memorable 1964 Olympics held in Tokyo. This design recalls on the images of the mark used for the Tokyo Olympics and incorporates the traditional crest motif signifying "first snow." The design as composed in such a way that it could be adapted for both vertical and horizontal use to facilitate display in all kinds of situations.
(Kazumasa Nagai)

札幌冬期オリンピックは日本で初めて開催された冬期オリンピックであり、1964年の東京オリンピックに続く一大イベントであった。東京オリンピックのイメージを継承しながら、日本らしい雪を表現するため、中央に日本の紋章の初雪をデザインして配した。このマークは色々な場所に展開しやすいように縦位置にも横位置にも使えるように考慮した。
(永井一正)

58b

SAPPORO'72

59a~59c
International Ocean Exposition, Okinawa '75 CI Plan
沖縄国際海洋博ーEXPO'75CI計画

Art Director & Designer
Kazumasa Nagai
永井一正
Client
Japan Association for The Ocean Exposition, Okinawa
財団法人沖縄国際海洋博覧会協会
1975

59a

As an exposition on the theme of the ocean and held in Japan, a mark was needed that would be both Japanese in flavor and yet universally understood. Using the rounded expression of waves found in the work of Hokusai and the Rinpa school artists, I adapted it with a modern stylized touch. I sought a dynamic image with the deep blue for the ocean and the bright blue for the sky separated by the white wavelets.
(Kazumasa Nagai)

沖縄海洋博は日本で開催される海洋をテーマにしたEXPOであるため、日本らしさと同時に世界に通用するマークを志向した。日本の伝統的な波頭の表現、例えば琳派や北斎などにも共通して見られる丸みを現代的にデザインしてみた。海の濃いブルーと空の明るいブルーの間に白い波頭が続く動きのあるイメージを出そうとした。
(永井一正)

59b

59c

60a～60b
World Design Exposition '89 VI Plan
世界デザイン博覧会VI計画

Art Director & Designer
Kazumasa Nagai
永井一正

Client
World Design Exposition Association
財団法人世界デザイン博覧会協会
1987

60a

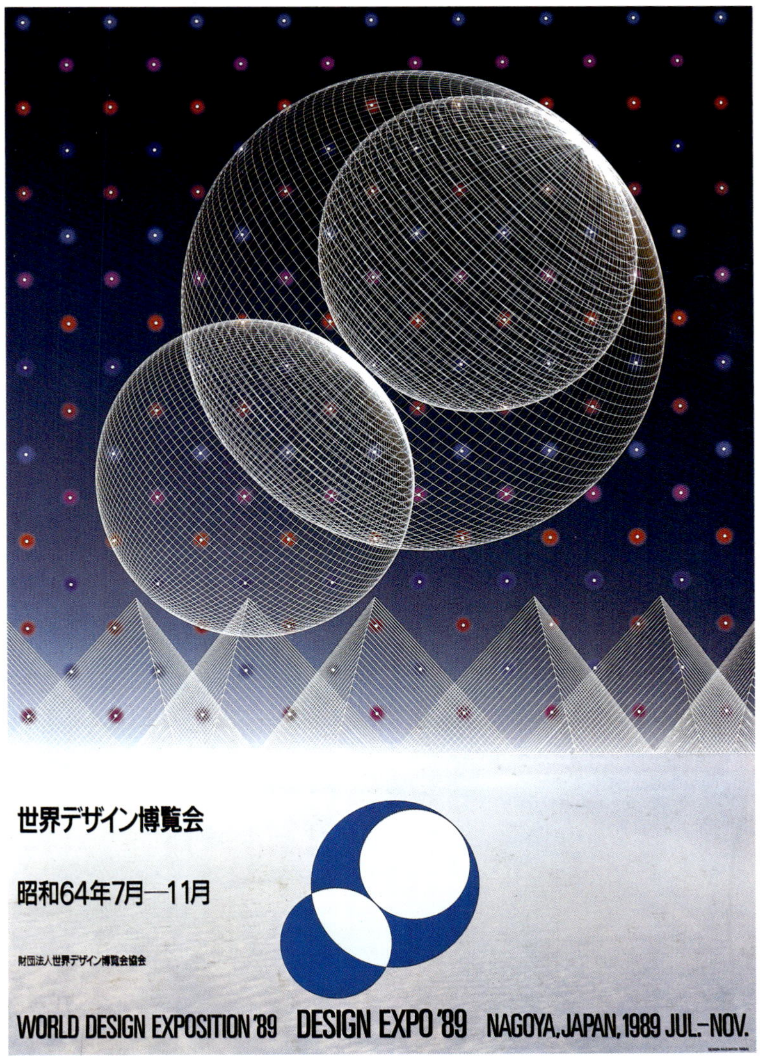

60b

The three circles symbolize the themes of the World Design Exposition '89—Humanity, Dream, Design—as well as the idea of ascending scale of human, globe, and the universe. This combination of simple circles is arranged in such a way as to symbolize man, the dreams he holds and nurtures, and the potential and hope that he can create through designing a rich culture on earth that will someday extend even to outer space.
(Kazumasa Nagai)

3つの円は〈世界デザイン博覧会〉のテーマである「人・夢・デザイン——都市が奏でるシンフォニー」のシンボライズであり、人、地球、宇宙というスケールの広がりでもある。このシンプルな円の組み合わせには、人類が夢を抱き、育み、デザインによって地上に文化を築き、やがては、未来宇宙へと飛び立つ可能性と祈りを込めた。
(永井一正)

61a〜61b
The Japan Peace Museum VI Plan
平和博物館VI計画

Art Director & Designer
Kazumasa Nagai
永井一正
Client
Association to Establish
The Japan Peace Museum
平和博物館を創る会
1984

61a

A movement has taken shape to build a "Peace Museum," a center of audio and visual materials dealing with the cause of peace built on civilian initiative. It is one of the projects of the Group for the Building of a Peace Museum, whose activities are dedicated to finding ways humankind can live in peaceful coexistence without nuclear weapons or war. This dove mark was created to express the prayer for peace.
(Kazumasa Nagai)

人類が核も戦争もない平和のうちに共存するためにはどうしたらよいのか、平和博物館を創る会はそのための様々な努力の試みの一つとして、市民の手で平和のための視聴覚資料センター平和博物館を建てる運動を呼びかけている。この鳩のマークはそんな平和への願いを込めて制作したものである。
(永井一正)

61b

62a～62c
The Ist International Triennial of Poster in Toyama, '85 VI Plan
第一回世界ポスタートリエンナーレ トヤマ '85VI計画

Art Director & Designer
Kazumasa Nagai
永井一正

Client
The Museum of Modern Art, Toyama
富山県立近代美術館
1985

62a

An international poster competition, held for the first time in Japan, was organized by the Toyama Prefectural Museum of Modern Art in 1985. An event long awaited among Japanese designers, it drew more than 1500 works by graphic artists from 37 countries. The mark was intended to symbolize the interchange among designers from all over the world.
(Kazumasa Nagai)

富山県立近代美術館の企画によって、日本で初めてポスターの世界的なコンペティションが行われた。これは日本のデザイン界待望のものであり、初回から1500点を超す作品が世界37ヶ国から集まった。このマークは世界のデザイナーの交流を意図して制作した。
(永井一正)

62b

62c

63a〜63e

Asahi Beer CI Plan

アサヒビールCI計画

Consulting
Dentsu Inc.
株式会社電通

Creative Director
Kazumasa Nagai
永井一正

Art Director
Masashi Uehara
上原 昌

Designer
Kahei Kiyono
清野嘉平
Yutaka Sasaki
佐々木豊

Client
Asahi Breweries, Ltd.
アサヒビール株式会社
1986

63a

"Live Asahi for Live People" is the new slogan created for the corporate identity program at Asahi Beer. The upward and forward diagonal lines of the brand name logotype created to match this slogan express vitality, dynamism and the distinctive flavor that only natural, genuine ingredients can assure. The corporate mark also bespeaks the company's readiness to take up any challenge and its orientation to future development. Bold blue, the corporate color, imparts a sense of freshness, exhilaration, and coolness.
(Kazumasa Nagai)

アサヒビールのCI導入による新しい経営理念から生まれた新しい心がLive Asahi for Live People である。これを表すコーポレートマークは、いきいきとした躍動感、自然で本物だけがもつ風格、さらに、たゆまぬ挑戦姿勢、将来への発展などの思いを込めてデザインしたものである。コーポレート・メインカラーのブルーはさわやかさ、すがすがしさ、清涼感を表している。
(永井一正)

63b

1F006
750ml 12缶入
ASAHI BREWERIES LIMITED
ASAHI
DRAFT BEER
ALL SEASONS TASTES GREAT
Asahi
ASAHI BEER FULLY CAPTURES THE TRUE TASTE OF NATURAL INGREDIENTS, AND ITS DISTINGUISHED ASAHI RICHNESS ADDS MORE TO YOUR ENJOYMENT.
ASAHI CAN BEER
1F008
350ml 24缶入
ASAHI BREWERIES LIMITED
ASAHI
DRAFT BEER
ALL SEASONS TASTES GREAT
1F011
ASAHI BEER CAPTURES THE TRUE TASTE OF NATURAL INGREDIENTS. DISTINGUISHED ASAHI RICHNESS ADDS MORE TO YOUR ENJOYMENT.
缶135ml
LIVE ASAHI FOR LIVE PEOPLE
Asahi
生
アサヒビール
2ℓ
1.65ℓ
1000ml
黒生
350ml
300ml
250ml
200ml
135ml
LIVE ASAHI FOR LIVE PEOPLE

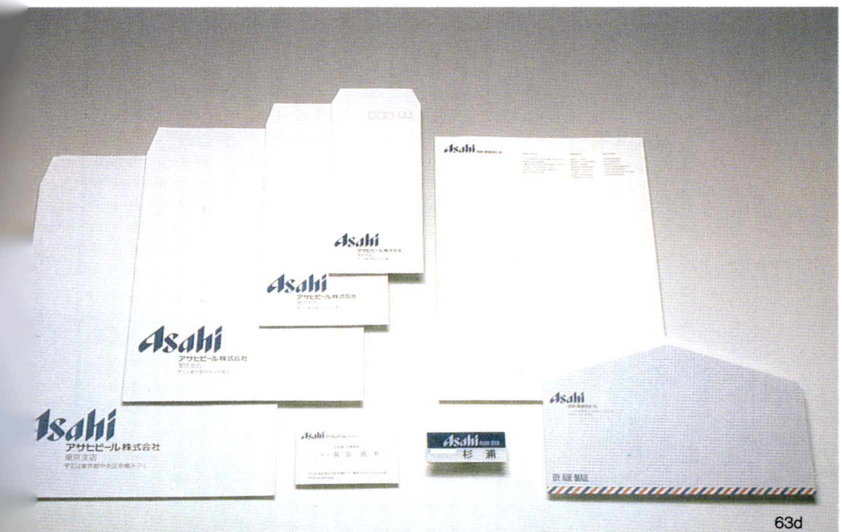
Asahi
アサヒビール株式会社
63d

Asahi
アサヒビール
63c

64

65

66

67

64
Minami-Nippon Broad-casting Symbol Mark
南日本放送 シンボルマーク
Art Director & Designer
Kazumasa Nagai
永井一正
Client
Minami-Nippon Broadcasting Co., Ltd. (MBC)
南日本放送株式会社
1983

65
Japan Amateur Sports Association Symbol Mark
日本体育協会 シンボルマーク
Art Director & Designer
Kazumasa Nagai
永井一正
Client
Japan Amateur Sports Association
財団法人日本体育協会
1974

66
Communications Tokyo Symbol Mark
コミュニケーション Tokyo シンボルマーク
Art Director & Designer
Kazumasa Nagai
永井一正
Client
Communications Industry Association of Japan
通信機械工業会
1984

67
Kawakichi Symbol Mark
カワキチ シンボルマーク
Art Director & Designer
Kazumasa Nagai
永井一正
Client
Kawakichi Co., Ltd.
株式会社カワキチ
1973

68

69

70

71

68
The Yasuda Trust and Banking Symbol Mark
安田信託銀行 シンボルマーク
Art Director & Designer
Kazumasa Nagai
永井一正
Client
The Yasuda Trust and Banking Co., Ltd.
安田信託銀行株式会社
1974

69
Governmental Publicity Communication Symbol Mark
政府広報 シンボルマーク
Art Director & Designer
Kazumasa Nagai
永井一正
Client
Prime Minister's Office
総理府
1978

70
Obihiro Shinkin Bank Symbol Mark
帯広信用金庫 シンボルマーク
Art Director & Designer
Kazumasa Nagai
永井一正
Client
Obihiro Shinkin Bank
帯広信用金庫
1976

71
Mitani Sangyo Symbol Mark
三谷産業 シンボルマーク
Art Director & Designer
Kazumasa Nagai
永井一正
Client
Mitani Sangyo Co., Ltd.
三谷産業株式会社
1983

72

73

74

72
Nico-Nico Nori Symbol Mark
ニコニコのり シンボルマーク
Art Director & Designer
Kazumasa Nagai
永井一正
Client
Nico-Nico Nori Inc.
ニコニコのり株式会社
1987

73
Konishiroku Symbol Mark
小西六 シンボルマーク
Art Director & Designer
Kazumasa Nagai
永井一正
Client
Konishiroku Photo Industry Co., Ltd.
小西六写真工業株式会社
1973

74
Kashiyama Symbol Mark
樫山 シンボルマーク
Art Director & Designer
Kazumasa Nagai
永井一正
Client
Kashiyama & Co., Ltd.
樫山株式会社
1977

75

76

77

78

75
Kaburagi Studio Symbol Mark
カブラギスタジオ シンボルマーク
Art Director & Designer
Kazumasa Nagai
永井一正
Client
Kaburagi Studio
株式会社カブラギスタジオ
1980

76
Ben General Symbol Mark
ベン・ゼネラル シンボルマーク
Art Director & Designer
Kazumasa Nagai
永井一正
Client
Ben General Corporation
ベン・ゼネラル株式会社
1980

77
Nichii Foods Symbol Mark
ニチイフーズ シンボルマーク
Art Director & Designer
Kazumasa Nagai
永井一正
Client
Nichii Foods Co., Ltd.
株式会社ニチイフーズ
1979

78
Ryoshoku Symbol Mark
リョーショク シンボルマーク
Art Director & Designer
Kazumasa Nagai
永井一正
Client
Ryoshoku Ltd.
株式会社リョーショク
1984

79

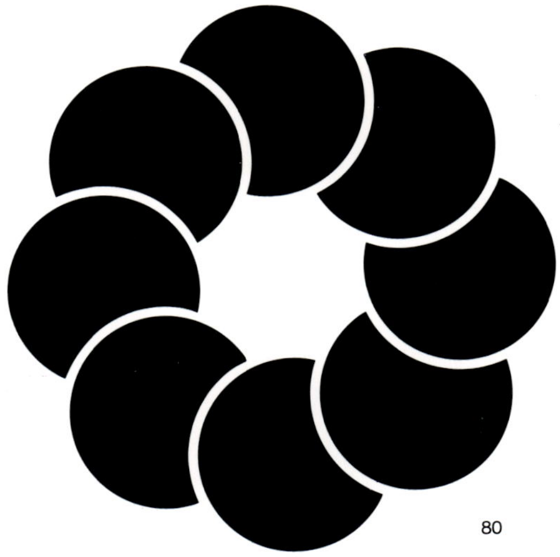

80

81

82

79
Tokyo Electric power company Symbol Mark
東京電力 シンボルマーク

Art Director & Designer
Kazumasa Nagai
永井一正

Client
The Tokyo Electric Power Company, Inc.
東京電力株式会社
1971

80
The Saitama Bank Symbol Mark
埼玉銀行 シンボルマーク

Art Director & Designer
Kazumasa Nagai
永井一正

Client
The Saitama Bank, Ltd.
株式会社埼玉銀行
1973

81
Nippon Lease Symbol Mark
日本リース シンボルマーク

Art Director & Designer
Kazumasa Nagai
永井一正

Client
Nippon Leasing Corporation
日本リース株式会社
1979

82
Daiya Symbol Mark
ダイヤ シンボルマーク

Art Director & Designer
Kazumasa Nagai
永井一正

Client
Daiya Co., Ltd.
ダイヤ株式会社
1975

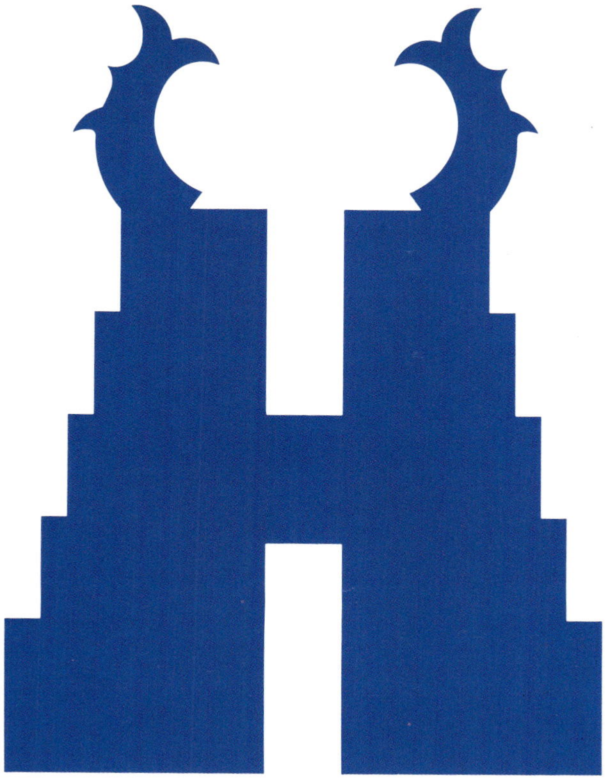

83

84

85

86

83
Himeji Municipality 100th Anniversary Symbol Mark
姫路市制100周年 シンボルマーク
Art Director & Designer
Kazumasa Nagai
永井一正
Client
Himeji Municipality
姫路市
1987

84
The University of the Air Foundation Symbol Mark
放送大学 シンボルマーク
Art Director & Designer
Kazumasa Nagai
永井一正
Client
The University of the Air Foundation
放送大学
1983

85
Japan Cultural Design Conference '83 Symbol Mark
日本文化デザイン会議'83 シンボルマーク
Art Director & Designer
Kazumasa Nagai
永井一正
Client
Japan Inter Design Conference
日本文化デザイン会議
1983

86
Nichii Symbol Mark
ニチイ シンボルマーク
Art Director & Designer
Kazumasa Nagai
永井一正
Client
Nichii Co., Ltd.
株式会社ニチイ
1979

87a〜87b
Sea & Islands EXPO, Hiroshima '89 VI Plan
'89海と島の博覧会ひろしまVI計画

Art Director & Designer
Shigeo Fukuda
福田繁雄
Client
Sea & Islands Exposition Association
財団法人'89海と島の博覧会協会
1987

87a

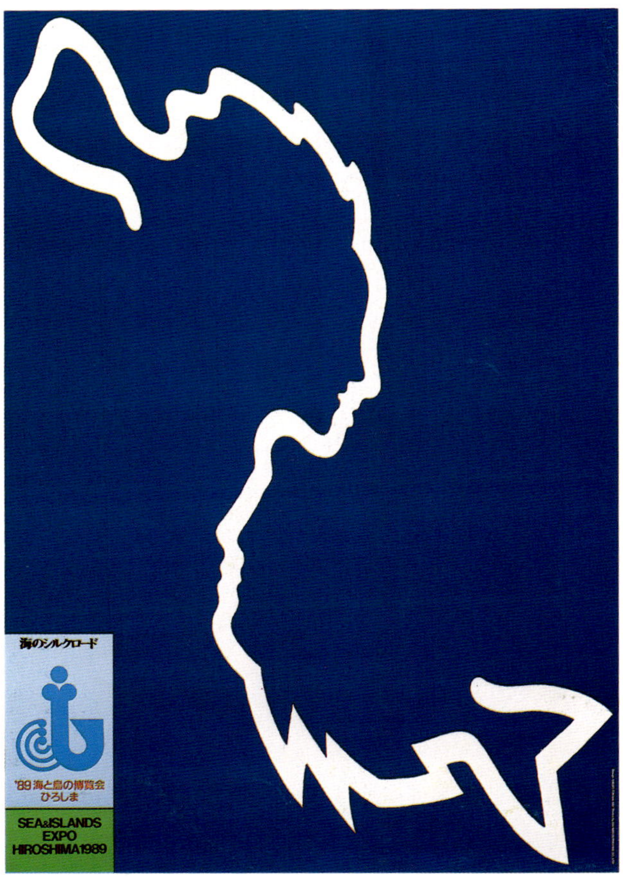

87b

One of our greatest pleasures in living in modern society is the hope of experiencing a kind of paradise of forest-covered islands floating in a pleasant sea. New meetings, new discoveries. This opportunity comes with the opening of the Sea & Islands Expo Hiroshima to be held in 1989. This vision is symbolized by the stylized form of the mermaid who cavorts in the seven seas.
(Shigeo Fukuda)

楽しさに満ちた楽園〈パラダイス〉を体験し、緑や水と親しみ、世界の島々を回遊する喜び。それは、現代社会に生きる私たちが最も期待する、新しいふれあいの発見です。そのプロローグが我が国で初めて開催される〈'89 海と島の博覧会・ひろしま〉です。その夢を七色の海に遊ぶ人魚に託して表現したものが、このシンボルマークです。
(福田繁雄)

88a〜88c
National Cultural Festival VI Plan
国民文化祭VI計画

Art Director & Designer
Shigeo Fukuda
福田繁雄

Designer
Masayoshi Nakajo — 88b
仲條正義
Koichi Sato — 88c
佐藤晃一

Client
Agency for Cultural Affairs,
Tokyo Metropolitan Government
文化庁・東京都
1986

88a

Culture is the product of human ingenuity; it is inextricably linked to an orderly daily life. The design seeks the image of ever-expanding culture in a jolly human figure, using mainly Murasaki purple, to evoke an image of brightness and a new future.
(Shigeo Fukuda)

文化は人間の知恵であり、秩序ある生活との結合であります。限りなく広がる文化へのあこがれを、歓喜の人形の構成で、イメージしたデザインです。日本古来の古代紫を基調に、明るさを加えて、新しい日本の未来色のイメージにした色彩計画です。
(福田繁雄)

88b

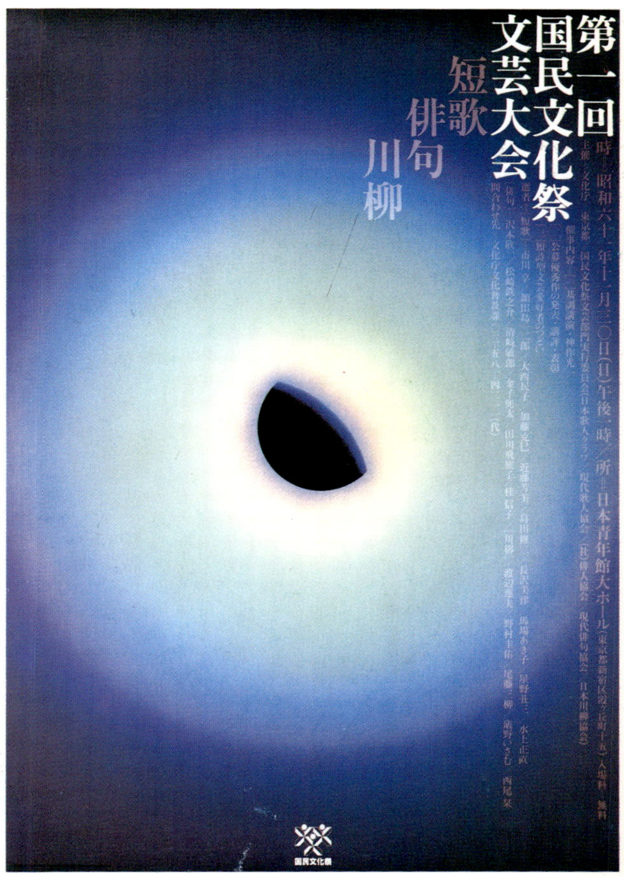

88c

89a~89d
National Children's Castle VI Plan
こどもの城VI計画

Art Director & Designer
Shigeo Fukuda
福田繁雄

Client
Child Welfare Foundation of Japan
財団法人日本児童手当協会
1985

This is indeed a castle built for children based on the ideal that they need both physical health and well-developed emotional and aesthetic sensitivities. It contains a theater, classrooms for music instruction and the plastic arts, play areas equipped with toys, exercise equipment for children and a pool as well as hotel facilities. The symbol mark expresses the vitality of a child born to face the future.
(Shigeo Fukuda)

健康と情操と感性のバランスをこどもたちに、という基本的な理念にもとづいて建設されたこどもの・ための文字通りのお城です。
劇場や音楽教室、造形教室など、遊具のある楽しい空間があり、こどものための運動施設やプール、さらに宿泊設備なども完備しています。元気なこどもが新しい未来に向かって生まれるイメージを視覚化したシンボルマークです。
(福田繁雄)

89b

89c

89d

90a〜90b
The Tokyo Tomin Bank CI Plan
東京都民銀行CI計画

Art Director & Designer
Shigeo Fukuda
福田繁雄

Client
The Tokyo Tomin Bank , Ltd.
株式会社 東京都民銀行
1976

90a

This mark for a suburban bank, which by its nature functions in a symbiotic relationship with local society, is composed of two superimposed elements: an arrow standing for the bank itself as an arena of local intercourse and evoking the human beings who gather there and the T of Tokyo Tomin Bank. The additional shape created by the overlapping of these two elements perfects the design. This plan was carried out by the Tokyo Tomin Bank Business Section Planning Group chiefly under the direction of the Space Co., Ltd. (Shigeo Fukuda)

「地域社会と共生する銀行」という基本的な理念にもとづいて、シンボルマークのエレメントを、TOKYO-TOMINの「T」と銀行店舗を地域の語らいの広場とし、そこに集まる人と人のイメージを矢印的な記号としてダブらせたデザインです。特定の形が重なることによって出来上がるもう一つの形をデザインのエレメントに簡潔させています。これらの企画制作は株式会社スペースを中心に、東京都民銀行業務部企画グループで進められたものです。
(福田繁雄)

90b

91a〜91b
The Hokkaido Takushoku Bank CI Plan
北海道拓殖銀行CI計画

Art Director & Designer
Shigeo Fukuda
福田繁雄

Designer
Kenji Ito
伊藤憲治

Client
The Hokkaido Takushoku Bank, Ltd.
株式会社北海道拓殖銀行
1969

91a

In creating this mark, the client asked that it be based on the motif of a star. The form of a star is already a stylized motif, so I determined to see how new and innovative I could be in rendering a star motif. I elongated its upward-pointing tip, changing it into an arrow-like shape. The enclosing circle makes it look also like a tree and highlights the progress-oriented stance of the bank.
(Shigeo Fukuda)

この銀行のマークの制作にあたっては、星をテーマとすることが要求された。星は、形そのものがすでにシンボル化されたものであるため、このマークのデザインの根本理念は、星をいかに新鮮にとらえるかというところに置かれた。そこで、星の頂点を上方に引き上げ、方向性をもった矢のように変形させ、さらにそれを円の中におさめることによって、星が樹木にも見え、また向上する意志にも見えるよう工夫した。
(福田繁雄)

91b

92a〜92b
Tennoji EXPO VI Plan
天王寺博VI計画

Art Director & Designer
Shigeo Fukuda
福田繁雄

Client
OSAKA 21ST CENTURY ASSOCIATION
財団法人大阪21世紀協会
1985

92a

92b

This logo was designed for an exposition held in Osaka's Tennoji Park in 1987, the slogan for which was "Inochi Ikiiki" [For Vitality in Life]. It seeks to evoke the image of people moving forward toward a bright future.
(Shigeo Fukuda)

大阪市天王寺公園で開催される博覧会のシンボルデザイン。スローガンは「いのちいきいき」。光に向かって前身する人間の未来をイメージしたデザイン。
(福田繁雄)

93a〜93c
Tenjin Core VI Plan
天神コアVI計画

Art Director & Designer
Shigeo Fukuda
福田繁雄

Client
Nishitetsu Tenjin Core Building Co., Ltd.
西鉄天神コアビル株式会社
1976

93a

This figure-ground reversable design integrates a T-shaped shape standing for "Tenjin" cut out of a circle and the silhouette profiles of two people evoking the arena of encounter and communication. Tenjin Core is a building housing a cluster of specialty shops. To match the fashionable interior decoration concept of the building, I sought a design that has fluid and youthful as well as clear-cut lines. The planning and production of this design program was carried out by the Nishitetsu Agency of Fukuoka.
(Shigeo Fukuda)

錯視図形のなかの多義図形の原理をベースに、シンボルマークがデザインされた。円形のなかのTENJINの頭文字「T」、その「T」を形づくる向き合った2人は、コミュニケーションと出合いの場をイメージしています。天神コアは、専門店を集合したショッピング・ビルでそのファッショナブルな店内構想から、明快な形態のなかにやさしい若さをとり入れようとしたデザインです。また、このプログラムの企画制作は、福岡の西鉄エージェンシーによって行なわれました。
(福田繁雄)

93b

93c

94a〜94d
Tsukuba EXPO'85 UCC Pavilion CI Plan
科学万博ーつくば'85 UCC館CI計画

Art Director & Designer
Shigeo Fukuda
福田繁雄
Client
Ueshima Coffee Co., Ltd.
上島珈琲株式会社
1984

94a

The coffee cup motif symbolizes the theme of the UCC displays: "Good Day, Nice Friends" (Coffee for Man at Home). The symbol mark was designed to express the hope that through the Science Expo '85, visitors could take home with them pleasant memories of their meeting with other people.
(Shigeo Fukuda)

UCCコーヒー館の出展テーマである"Good Day! Nice Friends! Coffee for Man at Home"をコーヒーカップにシンボライズしたものです。科学万博を通じて、人びとのふれあいと楽しい思い出をお持ち帰りいただく意図を表現したシンボルマークです。
(福田繁雄)

good day! nice friends!
94b

UCCコーヒー館
94c

good day! nice friends!
UCCコーヒー館
UCC
COFFEE
94d

95a〜95d
Yokohama Rehabilitation Center VI Plan
横浜市総合リハビリテーションセンターVI計画

Art Director & Designer
Shigeo Fukuda
福田繁雄

Client
Yokohama Rehabilitation Center
社会福祉法人横浜市リハビリテーション事業団
1987

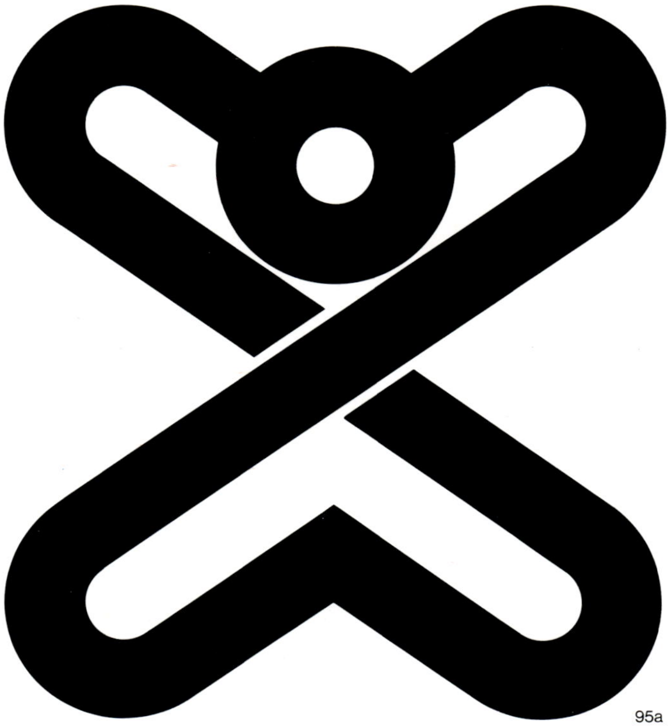

95a

I made somber tones and a sense of trustworthiness the basic guides for my design of the exterior signboard of this rehabilitation facility. The design makes the best of the thickness of the board, showing to pleasing effect the play of light and shadow in the symbol's curves. The execution for the second floor wall relief, in the endeavor to create a bright image for this kind of medical facility, is somewhat more colorful. The uninterrupted rotating line from which the compound symbol is embodied the concepts of "the will to live," "the medical profession," and "the hope of a better tomorrow." The relief has a paint finish. On the four walls of the center's multi-purpose hall are three-dimensional renderings in fiber reinforced plastic of the symbol mark scaled to the size of the space.
(Shigeo Fukuda.)

レリーフサイン：リハビリという施設の機能上、屋外サインには、色彩をおさえた格調ある信頼性を、基本デザインの指針としました。サイン板の厚みは、自然光による影の変化の効果を楽しめるように配慮しました。
2階フロアの壁面デザイン：医療施設を明るいイメージで統一するという、デザインの基本計画に合わせて、彩度を上げた色相で施工しました。シンボルマークの一本のラインによる循環性を、ここでは〈命ある意志と医学と明日への希望〉に連結させたデザインです。ペイント仕上げ。
多目的ホールの4ヶ所のコーナー上部にとりつけた壁面オブジェ：シンボルマークの循環性を、コーナーの空間に合せて立体的に表現したデザインです。素材はFRP製。
(福田繁雄)

95b

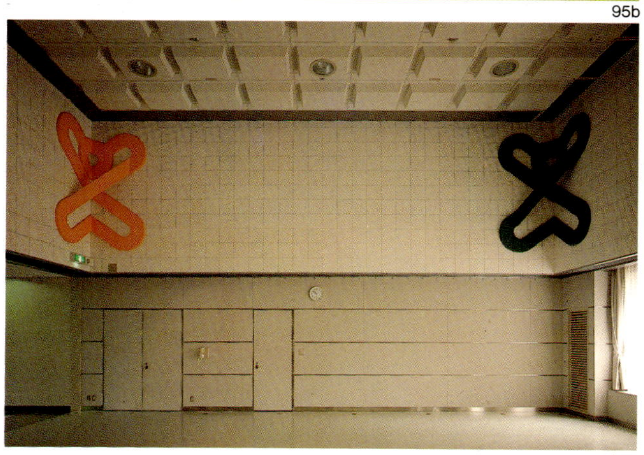

95c

95d

96a〜96e
Japan Stand TELECOM '87 VI Plan
テレコム'87日本スタンドVI計画

Consulting
SDG Co., Ltd.
株式会社エスデージー

Creative Director
Masatoshi Suzuki
鈴木正利

Art Director
Mamoru Ochiai
落合 鎮

Designer
Shigeo Fukuda
福田繁雄

Client
The Executive Council for TELECOM '87
テレコム'87実施協議会
1987

96a

For the main monument of the Japan pavilion of the Fifth World Electric Communications Exposition held in Geneva, Switzerland in 1987, we created a logo composed of ascending circles ascending as they might appear in animated action with the word "Japan" superimposed. The design is spread across a massive circular monument evoking the "rising sun" symbol of Japan.
(Shigeo Fukuda)

4年ごとにスイスのジュネーブで開催される第5回世界電気通信展示会（国際電気通信連合主催）の日本館のためのメインモニュメントは、アニメーションのように上昇する円と、JAPANのロゴで構成した。その巨大なモニュメントを平面に、日の丸のイメージの円の中にアレンジしたデザインです。
（福田繁雄）

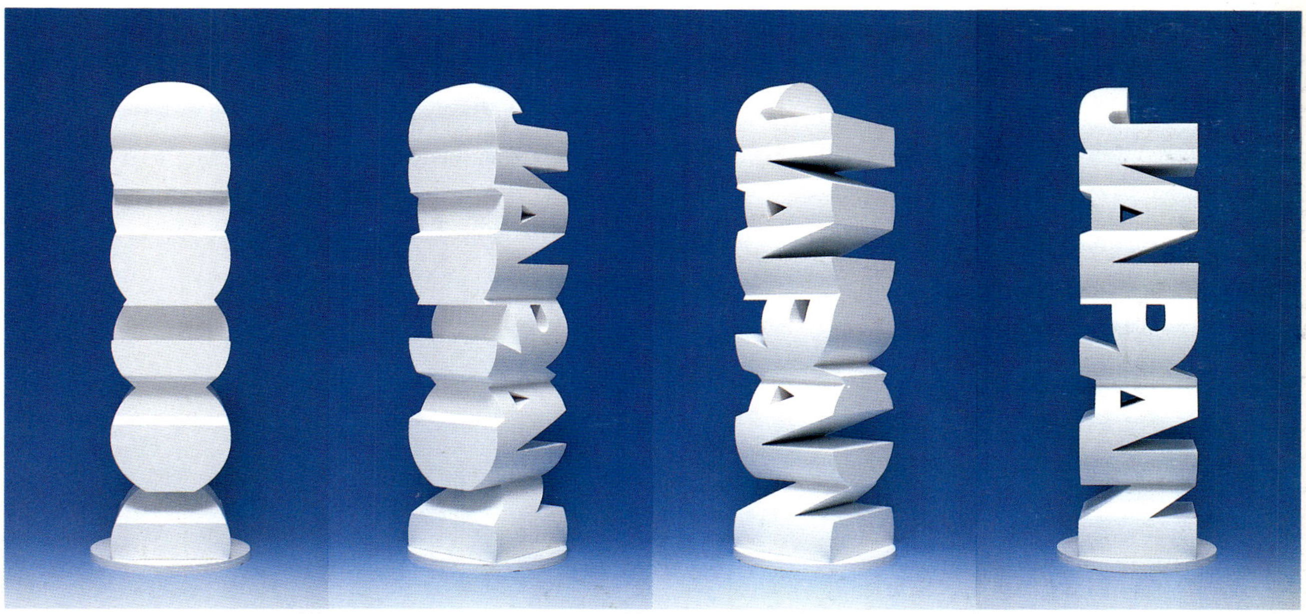

96b 96c 96d 96e

97
RISO KAGAKU Symbol Mark
理想化学工業 シンボルマーク

Art Director & Designer
Shigeo Fukuda
福田繁雄

Client
RISO KAGAKU CORPORATION
理想化学工業株式会社
1979

97

98
International Illustration Biennial Symbol Mark
国際イラストレーション・ビエンナーレ シンボルマーク

Art Director & Designer
Shigeo Fukuda
福田繁雄

Client
The Organization of International Biennial of Illustration, Tokyo (OIBIT)
東京イラストレーション・ビエンナーレ機構(OIBIT)
1985

98

99
Asahi Culture Center Symbol Mark
朝日カルチャーセンター シンボルマーク

Art Director & Designer
Shigeo Fukuda
福田繁雄

Client
Asahi Culture Center
朝日カルチャーセンター
1974

99

100

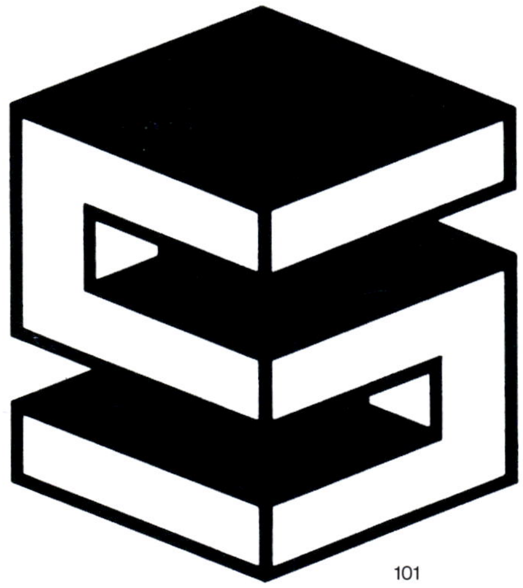

101

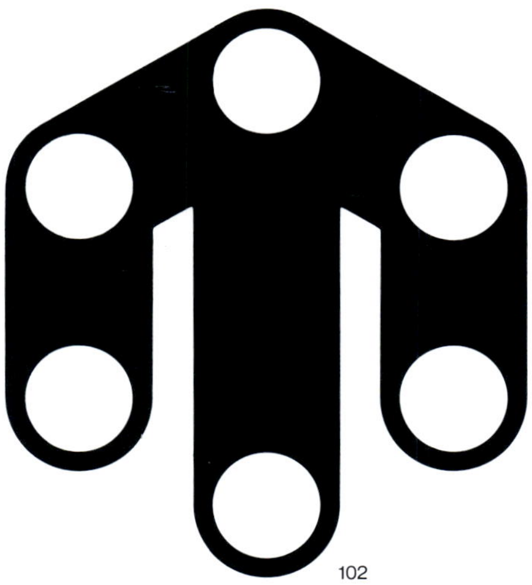

102

103

100
The 25th National Tree Planning Campaign Symbol Mark
第25回全国植樹祭 シンボルマーク

Art Director & Designer
Shigeo Fukuda
福田繁雄

Client
Iwate Prefecture Committee for the Promotion of Greenery
社団法人岩手県緑化推進委員会
1974

101
Gallery Seiho Symbol Mark
ギャラリーせいほう シンボルマーク

Art Director & Designer
Shigeo Fukuda
福田繁雄

Client
Seihosha Co., Ltd.
株式会社せいほう社
1976

102
Rikuyo-sha Symbol Mark
六耀社 シンボルマーク

Art Director & Designer
Shigeo Fukuda
福田繁雄

Client
Rikuyo-sha Publishing, Inc.
株式会社六耀社
1974

103
Competition of Gallery Atelier Nouveau Symbol Mark
アトリエ・ヌーボーコンペ シンボルマーク

Art Director & Designer
Shigeo Fukuda
福田繁雄

Client
The Seibu Department Stores, Ltd, Ikebukuro.
株式会社西武百貨店池袋店
1984

104a〜104f
Sanyo Scott BI Plan
山陽スコットBI計画

Art Director
Shin Matsunaga
松永　真
Client
Sanyo Scott Co., Ltd.
山陽スコット株式会社
1986

104a

104b

In commemoration of its 25th anniversary, Sanyo-Scott Company took the initiative in revamping its BI designs. A new design was selected in an international competition with a tissue paper box as the basic item, and it was expanded for use on all the firm's products as well. It led to a large scale BI plan. The chosen design possesses the main concept in the logo. Consequently, it also resulted in a new BI image. The use of the new logo with stripes in the background as a sub-pattern has been widened to include toilet papers, paper towels and other products.
(Shin Matsunaga)

山陽スコット社が創業25周年を記念して、世界に先駆けて全商品のデザインを一新した。ティッシュペーパーのボックスを代表テーマに国際コンペを行い、採用作を核に全商品のデザイン開発を計る、即ち規模の大きいBI計画であった。しかし決定作品はロゴタイプを基本エレメントとするものであり、従来のロゴタイプの見直しに最大のコンセプトを置くものであったために、必然的にCI計画にその意味を転じていくことになった。新シンボルロゴを中心にストライプがサブパターンとなってトイレットペーパー、ペーパータオルその他へ展開された。
(松永　真)

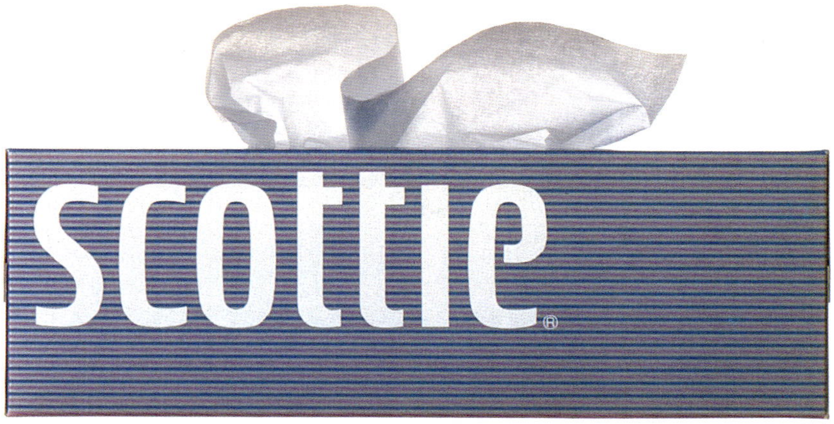

104c

104d

104e

104f

105a〜105d
Kirin Lemon 2101 BI Plan
キリンレモン2101 BI計画

Art Director & Designer
Shin Matsunaga
松永 真

Client
Kirin Brewery Co., Ltd.
キリンビール株式会社
1984

Developed by the Kirin Brewery Company to win back the young consumer, Kirin Lemon 2101 is a new variety of a well-known soft drink brand that has been on the market for 60 years. The design was conceived to make the product irresistible to junior and senior high school students. The figure "2101" is used for its design quality rather than for any meaning it may suggest. The slant of the lettering is aimed to enliven the design and the white cut-out effect on a plain blue ground expresses the fresh, bracing taste of the drink.
(Shin Matsunaga)

60年という長い歴史を背負ったキリンレモンの若者離れをとり戻すために開発されたサブブランドと言える。一にも二にも中高生ら自らが手を出したくなるデザインがコンセプトの全てであった。2101という数字が意味性を超えたデザイン要素として採用された。タイポグラフィーのかたまりを斜めに配して躍動感を、単色ブルー地に白ヌキで爽やかさを狙った。
(松永 真)

105a

105b

105c

105d

106a〜106d
Blendy BI Plan
ブレンディBI計画

Art Director & Designer
Shin Matsunaga
松永　真
Product Designer
Motomi Kawakami
川上元美
Advertising Design Creative Director
Takashi Nakahata
仲畑貴志
Client
Ajinomoto General Foods Inc.
味の素ゼネラルフーズ株式会社
1986

Blendy

106a

The logo and packaging design for AGF's new instant coffee. This design, from beginning to end, was to create an image of flavor that only coffee can offer. It is an instant coffee, and proud to be so. It evokes relaxation, a feeling of release from pressures and the contentment of family life with an air of confidence. The product sells very well. (Shin Matsunaga)

AGFのインスタントコーヒー、新ブレンディのBI計画である。最初から最後まで大切にしたことは、とにかくコーヒーがコーヒーらしくうまそうでありたかったことに尽きる。たかがインスタントコーヒー、されどインスタントコーヒーである。つまり、くつろぎと解放感、満ち足りた家庭の幸福の中で堂々とした存在感を獲得することであった。とにかくよく売れているらしい。
(松永　真)

106b

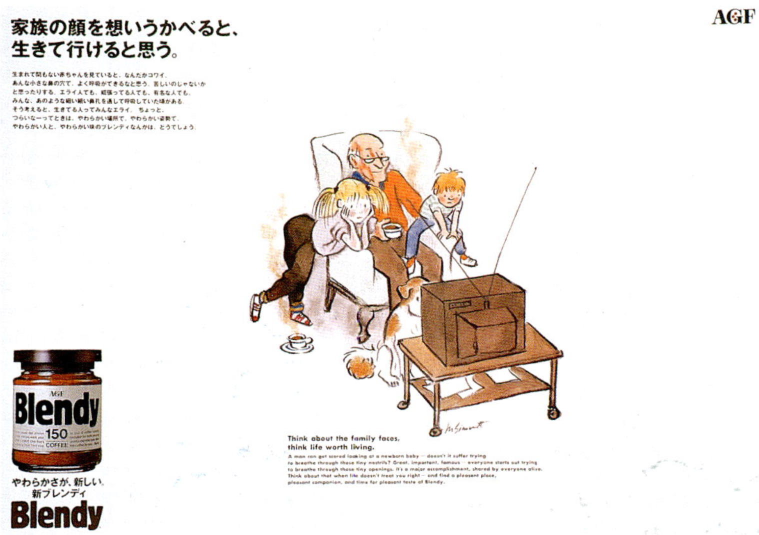

106c

106d

107a〜107c
Xanax CI Plan
ザナックスCI計画

Art Director & Designer
Shin Matsunaga
松永 真

Client
Xanax Corp.
株式会社ザナックス
1986

107a

This is the new logo created for Naniwa Co., Ltd., a general sports goods manufacturer when it changed its name to Xanax in 1986. The logo seeks an image of keenness and enterprising spirit and employs the trick of using bold letters for the first two characters to strengthen its impact even when used in reduced size. (Shin Matsunaga)

スポーツ用品の総合メーカーである株式会社ナニワが、1986年、Xanax（ザナックス）と改名した。その新シンボルロゴタイプである。シャープな進取性のイメージと共に、頭の2文字が強い形状になっており、動きの激しいスポーツ用品に展開する場合にも、フル形態と同様のインパクトが小スペースの中にも維持できるように、デザインされたものである。
（松永 真）

107b

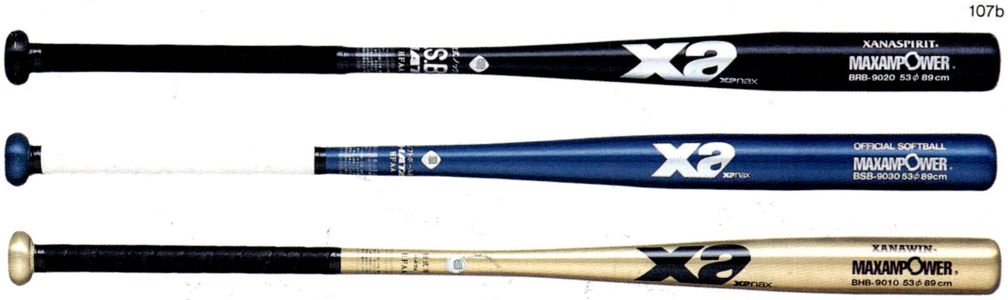

107c

Art Director & Designer
Shin Matsunaga
松永　真
Client
Tokorozawa Municipality
所沢市
1986

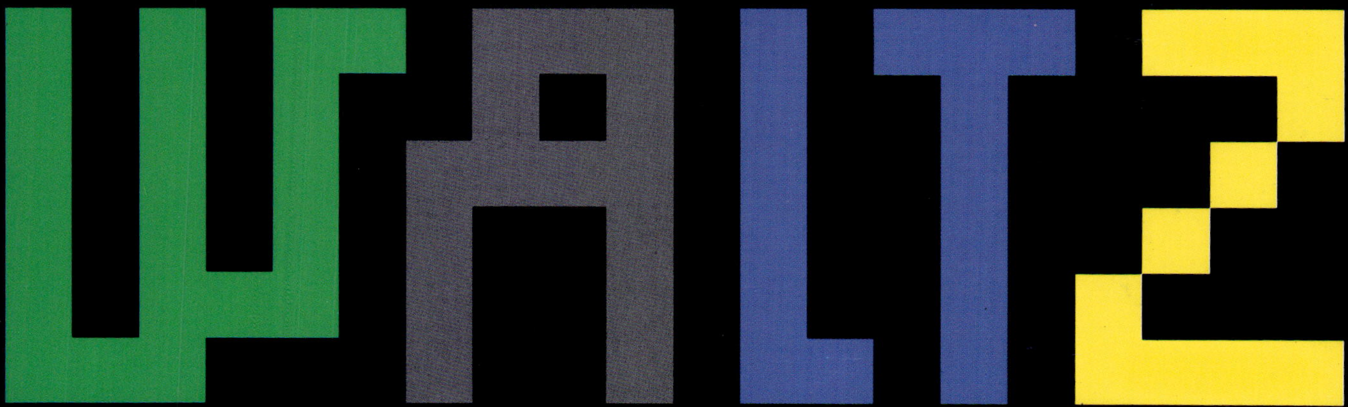

108a

This design was commissioned for the opening campaign of the Tokorozawa branch of Seibu Department Store which is part of the new station complex, named "Waltz," built in the suburb of Tokorozawa. Because the name embraces the entire complex, the typographic logo was treated 100% as a design element developed to adorn banners and shopping bags.
(Shin Matsunaga)

WALTZ（ワルツ）は、所沢駅前に建設された、所沢市の綜合ビルの名称である。百貨店"所沢西武"が、その中に同時に開店することが決まっており、オープニングキャンペーンと共に制作依頼を受けたものである。総称という意味合い上、今後一人歩きの少ないものであるので、オープニング時においては、ロゴそのものが100%のデザインエレメントとなって、大懸垂幕やショッピングバッグなどに展開された。
（松永　真）

108b

109a〜109c
BANDAI CI Plan
バンダイCI計画

Consulting
PAOS Inc.
株式会社パオス

Designer
Shin Matsunaga
松永　真

Client
Bandai Co., Ltd.
株式会社バンダイ
1983

At the time of Bandai's 30th anniversary, eight related companies in the toy and transportation industries were united under the Bandai name and a design competition was held to create a single logo for the entire group. In deciding on this mark, I was told there were great difficulties over interpretation of regulations because of the way it breaks up the company's name into two lines in the rectangular design.
(Shin Matsunaga)

バンダイが創立30周年を記念して玩具をはじめ運輸などの関連企業8社を統合して統一シンボルを制定することになり、デザインコンペが行われた。当マークが決定するには、社名が矩形の中で2行になっていることで法規解釈上大変な難産であったと聞いている。
(松永　真)

109a

109b

109c

110a〜110c
Hankyu Department Store VI Plan
阪急百貨店VI計画

Art Director & Designer & Illustrator
Shin Matsunaga
松永 真

Client
Hankyu Department Store Inc.
株式会社阪急百貨店
1982

Hankyu Department Store in Umeda, Osaka decided to renew the designs for its promotional materials. The theme was to be "life" and "nature," and I drew a group of designs based on traditional nature-related themes. The keynote was suitability for particular sales departments as well as for use as an element uniting the store as a whole; it was also to incorporate several thematic colors combined in a variety of ways. Unfortunately, only one of the designs was ultimately used, but Hankyu did draw on the basic concept when it launched its branch store in Yurakucho, Tokyo.
(Shin Matsunaga)

梅田阪急が販促物のデザイン再開発をすることになった。生命や自然をテーマにして花鳥風月などあらゆるモチーフで数多くのドローイングを試みた。そして豊富なテーマカラーの中で色々な組み合わせをしながら、売場適性に応じた展開と店全体としての音響効果のような広がりを見せたいというのが主眼であった。残念ながら採用されたのは一つのモチーフにとどまったが、有楽町阪急の進出で本来のコンセプトの一部が汲み上げられることになったのである。
(松永 真)

110a

110b

110c

NEWS

111a

111b

111a〜111k
NEWS BI Plan
ニューズBI計画

Art Director & Designer
Shin Matsunaga
松永　真
Product Designer
Motomi Kawakami
川上元美
Client
Kirin-Seagram Ltd.
キリン・シーグラム株式会社
1983

This is the logotype and packaging designed for promotion of Kirin-Seagram Ltd.'s new whisky brand "News," developed for the tastes of the so-called Dankai generation (those of the postwar baby-boom who are now in their mid-thirties). We sought an archetypical image of whisky to suit contemporary life; the unprepossessing presentation we came up with was so novel that it sent shock waves through the whole whisky industry, sparking an unprecedented bottle design battle.
(Shin Matsunaga)

団塊の世代に標準を合わせて開発されたキリンシーグラム社の新ウイスキーNEWSのBI計画である。制作者のコンセプトは、あくまで現代の生活空間におけるウイスキーの原型探しであった。結果導き出した何でもないはずの形状が過激な差別化として大きな反響をまき起こし、結果的には新ボトル戦争の火つけ役となった。
(松永　真)

111c

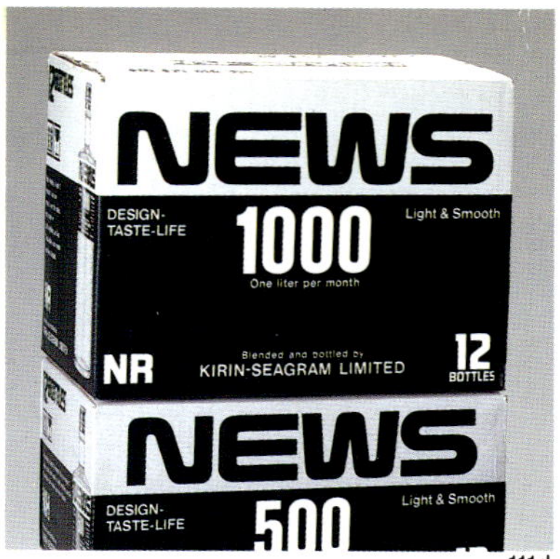

111d

111e

111f

111g

111h

111i

111j

111k

KAGOME 112

MYCITY 113

SORD 114

112
Kagome Logotype
カゴメ ロゴタイプ

Art Director & Designer
Shin Matsunaga
松永　真

Client
Kagome Co., Ltd.
株式会社カゴメ
1983

113
My City Logotype
マイ シティ ロゴタイプ

Creative Director
Yoshihiro Iwanaga
岩永嘉弘

Art Director & Designer
Shin Matsunaga
松永　真

Client
Shinjuku Station Building Co.
株式会社新宿ステーションビルディング
1978

114
Sord Logotype
ソード ロゴタイプ

Art Director & Designer
Shin Matsunaga
松永　真

Client
Sord Computer Corporation
ソード株式会社
1983

SEIYU 115

Autorama 116

Kaneka 117

115
Seiyu Logotype
西友 ロゴタイプ

Creative Director
Ikko Tanaka
田中一光

Art Director & Designer
Shin Matsunaga
松永 真

Client
The Seiyu Ltd.
株式会社西友
1983

116
Autorama Logotype
オートラマ ロゴタイプ

Art Director & Designer
Shin Matsunaga
松永 真

Client
Autorama Inc.
株式会社オートラマ
1981

117
Kaneka Logotype
カネカ ロゴタイプ

Art Director & Designer
Shin Matsunaga
松永 真

Client
Kanegafuchi Chemical Industry Co., Ltd.
鐘淵化学工業株式会社
1985

118

119

120

121

122

123

118
Ings Logotype
イングス ロゴタイプ
Art Director & Designer
Shin Matsunaga
松永 真
Client
Hankyu Department Store, Inc.
株式会社阪急百貨店
1982

119
Ube Logotype
宇部興産 ロゴタイプ
Art Director & Designer
Shin Matsunaga
松永 真
Client
Ube Industry Limited
宇部興産株式会社
1985

120
LEE Logotype
リー ロゴタイプ
Art Director & Designer
Shin Matsunaga
松永 真
Client
Shueisha Publishing Co., Ltd.
株式会社集英社
1983

121
Select Logotype
セレクト ロゴタイプ
Art Director & Designer
Shin Matsunaga
松永 真
Client
Shiseido Co., Ltd.
株式会社資生堂
1984

122
Savas Logotype
サバス ロゴタイプ
Art Director & Designer
Shin Matsunaga
松永 真
Client
Meiji Seika Kaisha, Ltd.
明治製菓株式会社
1979

123
More Logotype
モア ロゴタイプ
Art Director & Designer
Shin Matsunaga
松永 真
Client
Shueisha Publishing Co., Ltd.
株式会社集英社
1977

124

125

126

127

128

129

124
Escargot Books Symbol Mark
エスカルゴ ブックス シンボルマーク
Art Director & Designer
Shin Matsunaga
松永 真
Client
Nihon-Jitsugyo Publishing Co., Ltd.
株式会社日本実業出版社
1982

125
CBS-Sony Books Symbol Mark
CBS ソニー ブックス シンボルマーク
Art Director & Designer
Shin Matsunaga
松永 真
Client
CBS-Sony Publishing Co., Ltd.
株式会社CBS ソニー出版
1979

126
Honma Golf Symbol Mark
本間ゴルフ シンボルマーク
Art Director & Designer
Shin Matsunaga
松永 真
Client
Honma Golf Co., Ltd.
本間ゴルフ株式会社
1985

127
Sankei Bunko Symbol Mark
サンケイ文庫 シンボルマーク
Art Director & Designer
Shin Matsunaga
松永 真
Client
Sankei Publishing, Inc.
株式会社サンケイ出版
1986

128
Asahico Symbol Mark
アサヒコ シンボルマーク
Art Director & Designer
Shin Matsunaga
松永 真
Client
Asahi Industries Co., Ltd.
朝日工業株式会社
1986

129
Nippon Paint Symbol Mark
日本ペイント シンボルマーク
Art Director & Designer
Shin Matsunaga
松永 真
Client
Nippon Paint Co., Ltd.
日本ペイント株式会社
1983

130a～130g
KENWOOD CI Plan
ケンウッドCI計画

Consulting
PAOS, Inc.
株式会社パオス

Designer
PAOS, Inc.
株式会社パオス

Client
Kenwood Corporation
株式会社ケンウッド
1982

KENWOOD

130a

130b

Kenwood made a major changeover in its management the occasion for adopting a whole new corporate image strategy. It stopped use of the former double brand, Trio/Kenwood, unifying all its products under the Kenwood brand. The logo is intended to embody "quality," "state-of-the-art" technology and "incisiveness." The pointed form of the letter "W" with the triangle mark and the triangle formed of arrow-like stripes suggests the broad growth potential of the company.
(Motoo Nakanishi)

ケンウッドのCIシステムは、経営者の交替を機にイメージ一新の戦略として実施され、それまでのTRIO, KENWOODのコーポレートブランド併用から、KENWOODブランドに一本化していくことが決定された。デザインは高品質、先進性、鋭さというキーワードを象徴化したもので、Wの文字に視覚的なポイントとしての三角形を置き、さらに変形ストライプの三角形のグラフィックキャラクターによって、幅広い展開が可能になっている。
(中西元男)

130c

130d

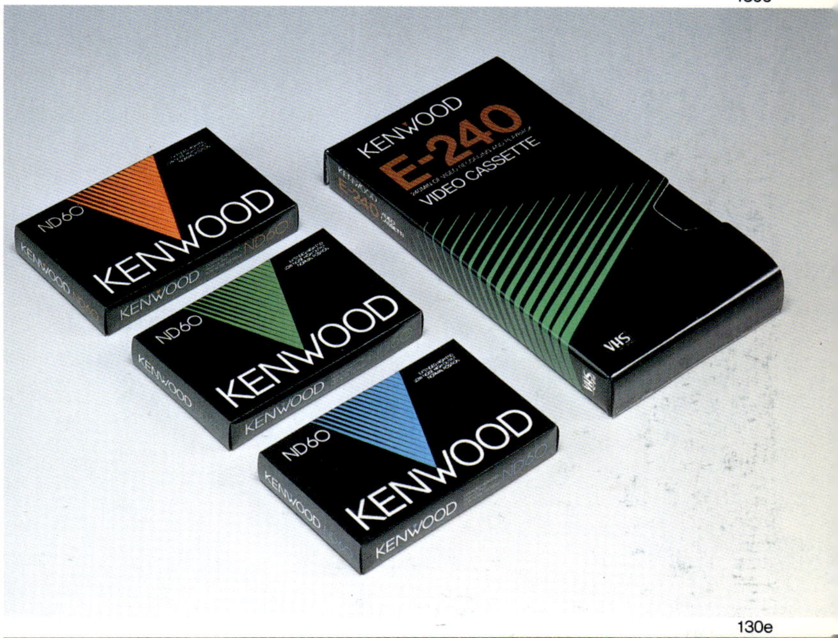

130e

130f

130g

131a〜131g
BRIDGESTONE CI Plan
ブリヂストンCI計画

Consulting
PAOS, Inc.
株式会社パオス

Designer
PAOS, Inc.
株式会社パオス

Client
Bridgestone Corporation
株式会社ブリヂストン
1984

131a

131b

131c

Bridgestone prides itself on its overwhelming share of the domestic tire market. In 1981, the 50th anniversary of its founding, the company began to work on the introduction of a new corporate image and changed its formal name from Bridgestone Tire Co., Ltd. to Bridgestone Corporation. The new CI symbolizes the company's goal to play an increasing social role by expanding the territory of its enterprise throughout the world. It is based on the concepts of advancement, internationality, and reliability.
(Motoo Nakanishi)

ブリヂストンは国内で圧倒的シェアを誇るタイヤメーカーである。創業50周年 (1981年) を機にCI導入の検討がなされ、社名もブリヂストンタイヤ株式会社から株式会社ブリヂストンへと変更された。ブリヂストンのCIは、世界に向かって事業領域を拡大する社会的役割を果たす企業を目指し、先進性、国際性、信頼性をコンセプトとしてデザインされている。
(中西元男)

131d

131e

131f

131g

132a〜132e
Kirin Beer CI Plan
キリンビールCI計画

Consulting
PAOS, Inc.
株式会社パオス

Designer
PAOS, Inc.
株式会社パオス

Client
Kirin Brewery Co., Ltd.
キリンビール株式会社
1984

132a

132b

For the new Kirin logo, the typographical part of the logo has been separated from the graphic representation of the mythological "Kirin" the company mascot creature. The new corporate concept stresses the departure from a symbol used solely on the label of beer bottles and cans to one suitable to its active efforts to branch into new areas of enterprise. Refinements in the outline of the Kirin, moreover, have been made in order to strengthen its impression on the subconscious of the consumer. (Motoo Nakanishi)

新シンボルは、ビールのラベルに源をもつシンボル構造から、新事業分野を積極的に展望した新企業理念の下で、KIRINロゴを霊獣キリンから切り離した点に最大の意味がある。一方、ビール事業のシンボルである霊獣の造型は、消費者の深層心理への投影を意図して超緻密にリファインされている。(中西元男)

132c

132d

KIRIN BREWERY COMPANY, LIMITED
DRAFT BEER
KIRIN
キリン生ビール
2ℓ
3ℓ
キリンビール
BREWED FOR GOOD TIMES
1.2ℓ
1957ml
KIRIN BEER
BLACK
350ml
KIRIN
STOUT
IMPORTED
KIRIN BEER
33 cl
LAGER BEER
BREWED IN JAPAN
こきりん
500ml
1000ml
350ml
BEER

133a〜133e
INAX CI Plan
イナックスCI計画

Consulting
PAOS, Inc.
株式会社パオス

Designer
PAOS, Inc.
株式会社パオス

Client
INAX Corporation
株式会社INAX
1985

INAX

133a

The adoption of the INAX corporate image for this manufacturer of tile and other housing and building materials constituted more than a simply change of company name from Ina Ceramics to INAX; it was an act symbolic of a macro-level corporate challenge to the coming age. The company has not only developed and introduced a thoroughly new visual image, but instituted various improvements in the corporate setup, developed new lines of business, and launched a broad range of corporate image activities. The "space symbol" that is the core of the INAX CI is based on the company motto, "creating and offering to customers attractive environments," and gives graphic form to the content of the space design business in which the company engages.
(Motoo Nakanishi)

タイル・住宅設備機器の総合メーカーINAXのCIは、伊奈製陶からINAXへの単なる社名変更の域を超えた、次の時代に向かって挑戦するマクロ的な企業姿勢の象徴行為である。企業体質の改善、新事業の開発等、幅広いCI活動が展開されている。CIシステムの核となるシンボル〈スペースシンボル〉は、環境美の創造、提供を企業理念として、スペースデザインをビジネスとするINAXの業容を表現している。
(中西元男)

133b

133c

133d

133e

134a~134b
XSITE VI Plan
エクサイトVI計画

Consulting
PAOS, Inc.
株式会社パオス

Designer
PAOS, Inc.
株式会社パオス

Client
INAX Corporation
(XSITE Division)
株式会社INAX
(エクサイト事業部)
1986

134a

The name of this enterprise, combining the "X" of the parent company INAX with the word "site" where its activities are materialized, is the distillation of corporate strategy. The graphic symbol consisting of a human figure reclining in a water droplet symbolizes the image of comfort and relaxation in toilet and bathroom spaces and softens the image of the logo as a whole. (Motoo Nakanishi)

INAXが、事業開発型CIの尖兵として送り出したXSITE（エクサイト）事業は、INAXのキーワードXと、その具現の場SITEを組み合わせたネーミングに戦略的意図が凝縮している。一方水滴の中でなごむ人間をモチーフとしたキャラクターマークが、トイレ、バスという第3の空間に求められる心地よいゆらぎイメージを象徴し、ブランドロゴにやわらかな表情を与えている。
(中西元男)

134b

135a〜135c
The Tokio Marine And Fire Insurance CI Plan
東京海上火災保険CI計画

Consulting
PAOS, Inc.
株式会社パオス

Designer
Chermayeff & Geismar Associates
シャマイエフ アンド ガイスマー アソシエイツ
PAOS, Inc.
株式会社パオス

Client
The Tokio Marine And Fire Insurance Co., Ltd.
東京海上火災保険株式会社
1984

The Tokio Marine and Fire Insurance Company, founded more than 100 years ago, has the longest history of any nonlife insurance company in this country. A completely new image directed at the general consumer was required as the company began to move from marine insurance into the field of household insurance and in view of the advent of the high-tech information society and the new rivalry in the financial market. The new logo is based on the basic concepts of the corporate image; "familiarity," "activism," and "foresight." (Motoo Nakanishi)

東京海上は、創業100年を超えるわが国で最も伝統のある最大の損害保険会社である。かつての海上保険中心から家計保険分野への移行、高度情報化社会、金融新競合時代の到来という中で、主に一般消費者に向けてイメージ一新が必要とされていた。新しいシンボルは親しみ、行動力、先見性を基本イメージとしてデザインされたものである。
(中西元男)

TOKIO
MARINE

135a

135b

135c

136a～136d
Zojirushi CI Plan
象印マホービンCI計画

Consulting
PAOS, Inc.
株式会社パオス
Dentsu Inc.
株式会社電通

Designer
PAOS, Inc.
株式会社パオス

Client
Zojirushi Corporation
象印マホービン株式会社
1986

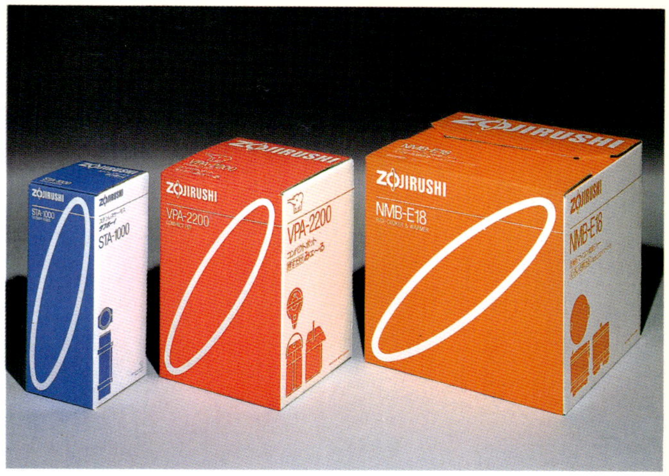

136a

136b

136c

136d

This design was developed to solidify the image of a manufacturer of household goods and appliances utilizing the latest technology. The red ring encircling the "O" of ZO symbolizes the up-to-date know-how, advanced technology, and internationalism at which Zojirushi aims and at the same time expresses the company's vigorous efforts to foster a humanistic bond between the consumer and industry.
(Motoo Nakanish)

先進技術を核とする総合生活用品メーカーとしてのイメージ確立を狙いとしてデザイン開発がなされた。ZOにかかる赤い環〈インターリンク〉で、象印の目指す先進性、技術力、国際性をシンボライズすると同時に、人間と企業をあたたかく結ぶ躍動感溢れる企業姿勢を表現している。
(中西元男)

137a～137f
Toray Industries CI Plan
東レCI計画

Consulting
PAOS, Inc.
株式会社パオス

Designer
Colin Forbes (Pentagram Design Ltd.)
コリン・フォーブス(ペンタグラム)
PAOS, Inc.
株式会社パオス

Client
Toray Industries, Inc.
東レ株式会社
1986

'TORAY'

137a

'TORAY'

137b

'TORAY'

137c

The symbol of Toray's corporate identity was made deliberately neutral, readable and yet distinctive so that it would be suitable for use in any of the diverse enterprises into which the company may extend its activities from now on. The single quotation marks are used as they are to label dialogue and to give stress to a word. They express the company's desire to be a concern that is in constant communication with society and where there is plentiful internal dialogue, and also to be a company whose presence is outstanding in society.
(Motoo Nakanishi)

東レのコーポレートシンボルは、今後予想される極めて多様な事業分野に対し幅広い対応が可能なように、ニュートラルで読みやすい中にも独特の個性を求めてデザインされた。両端に配されたコーテーションマークは、一般に対話を意味すると同時に文章中の特定の単語を際立たせる時に使われる。社会と対話し、社内に対話のあふれている企業であるとともに、社会の中で際立った存在でありたいという企業姿勢を表現している。
(中西元男)

137d

137e

137f

138a〜138c
Mitsui Real Estate Sales CI Plan
三井不動産販売CI計画

Consulting
PAOS, Inc.
株式会社パオス

Designer
PAOS, Inc.
株式会社パオス

Client
Mitsui Real Estate Sales Co., Ltd.
三井不動産販売株式会社
1981

138a

138b

138c

Introducing a new corporate concept to the Mitsui Real Estate Sales distribution firm, the Mitsui Real Estate Sales Housing Network was renamed "Mitsui Rehouse." The features of the CI system are designed to foster a sense of friendliness and trust and symbolize improved living through purchase of new dwellings. After adoption of the new logo, the word "rehouse" became synonymous with the general term for buying new homes and Mitsui's real estate sales increased sharply.
(Motoo Nakanishi)

三井不動産販売の流通事業の名称〈三井不動産販売住まいのネットワーク〉を新事業概念の導入によって、〈三井のリハウス〉に変更。そのデザインシステムの特徴は、住みかえによる生活上昇の表現、親しみと信頼感の醸成であった。導入後リハウスの名称は住みかえの代名詞として定着し、三井のリハウスの業績も急上昇した。
(中西元男)

139a～139c
The Sumitomo Bank CI Plan
住友銀行 CI計画

Consulting
PAOS, Inc.
株式会社パオス
Designer
PAOS, Inc.
株式会社パオス
Client
The Sumitomo Bank, Ltd.
株式会社住友銀行
1986

住友銀行
SUMITOMO BANK

139a

Even among the handful of major city banks in Japan, Sumitomo is noted for its high earnings and reliability. With the advent of the "new age in finance," the company began in 1986 to develop a new design system, one of the goals of which is to refurbish and up-date the visual environment of its branch banks, the central arenas of its business. The new graphics for the bank's name and other corporate expressions were to be based on an image of "newness based on tradition. Their introduction was timed to coincide with the merger of Sumitomo and the Heiwa Sogo Bank which took place in October that year, and deposit books and bank cards with the new designs were issued. The work of developing a new atmosphere for banking business is still going on.
(Motoo Nakanishi)

住友銀行は、数あるわが国都市銀行の中でも、収益性と信頼性の高さでつとに著名である。新金融時代を背景として、銀行ビジネスの最重要拠点である店舗における視覚環境の整備、一新をひとつの目標に、1986年初めより、新しいデザインシステムの開発がなされた。〈伝統が生きる新しさ〉を基本イメージとした行名書体をはじめとする新デザインは、同年10月平和相互銀行合併を機に、旧平和相互銀行店舗のサインから導入され、同時に通帳やカードなどのデザインも一新されていった。現在も、新しいバンクビジネス空間の構築を目指して開発作業は継続中である。
(中西元男)

139b

139c

140a〜140e

Yamato CI Plan

やまと CI計画

Consulting
PAOS, Inc.
株式会社パオス

Designer
PAOS, Inc.
株式会社パオス

Client
Yamato Co., Ltd.
株式会社やまと
1986

140a

The Yamato chain of Kimono stores sought to revitalize its business by creating a new image focussing mainly on its shops. The idea was to design a new visual environment, replacing the old image of the traditional Japanese kimono shop with a new, up-to-date one that might be described as "kimono Japonesque." A new corporate symbol and new shop designs were developed that combine Japanese and Western motifs. The scheme was implemented mainly for the fashionable shops frequented by relatively young customers.
(Motoo Nakanishi)

着物のやまとは、チェーン店の活性化をねらいとして、店舗を中心とする新しい環境イメージづくりに取り組んだ。旧来の和装着物イメージから、キモノジャパネスクとも呼べる新イメージの創出がテーマとなり、和と洋を感覚的に融合させた新シンボル及び新店舗デザインが開発された。新デザインは、基本デザイン要素から店舗及び什器、販促物、ユニフォーム、伝票、そしてショーイングに至るまで、実験導入という形式をとりながら徹底した試みがなされ、特に比較的若年層顧客をターゲットとした尖端型立地店を中心に展開されていった。
(中西元男)

140b

140c

140d

140e

141a〜141f
Suntory Hall VI Pian
サントリーホールVI計画

Art Director
Takenobu Igarashi
五十嵐威暢
Shigeshi Omori
大森重志

Designer
Takenobu Igarashi
五十嵐威暢
Yukimi Sasago
笹子行美
Debi Shimamoto
デビ・島本
Shizuko Ushijima
牛島志津子

Client
Suntory Ltd.
サントリー株式会社
1986

SUNTORY HALL

141a

In 1983, Suntory Ltd. celebrated two red-letter anniversaries-20 years of beer brewing and 60 years in the whisky business. Construction of its own concert hall, a long-cherished dream, was begun that year. The project was undertaken with the vision of making it the world's most acoustically perfect concert hall, and its identifying logo was developed on the theme of "echo." The design was ultimately created from the Chinese character "hibiki," or "kyo," meaning "echo, reverberation." The logo was also effective when produced as a sculpture, for signs, advertising, promotional materials and merchandise.
(Takenobu Igarashi)

1983年、サントリー株式会社はウイスキー作り60年、ビール作り20年を記念して、かねてからの念願であったコンサートホールの建設に着手した。この事業は世界一響きの美しいホールをという高い理想のもとに進められ、シンボルマークの開発は響きをテーマに繰り広げられた。最終的には響の文字そのものをモチーフとしたデザインが採用され、これは彫刻サイン、広告宣伝物、グッズ等に効果的に展開されている。
(五十嵐威暢)

141b

141c

141d

141e

141f

142a〜142b
Mitsui Bank CI Plan
三井銀行CI計画

Consulting
Total Media Development Institute Co., Ltd.
トータルメディア開発研究所
Communication Project Inc.
株式会社コミュニケーションプロジェクト
Toppan Printing Co.
凸版印刷株式会社

Art Director
Takenobu Igarashi
五十嵐威暢

Designer
Takenobu Igarashi
五十嵐威暢
Yukimi Sasago
笹子行美

Client
The Mitsui Bank, Limited
株式会社三井銀行
1984

三井銀行

142a

The Mitsui Bank, Ltd. introduced a new CI scheme in 1983 as part of activities marking the 300th year since its founding. The wing motif combines the harmony and friendliness evoked by a circle with the solidity and dependability of the square to create an impression of warmth and stability. The three oblong shapes darting diagonally across the circle are symbolic of growth and vitality. Mitsui's corporate color, a fresh light blue, expresses "stability," "vitality," "approachability," and "modernity."
(Takenobu Igarashi)

1983年、三井銀行は創業300年の一環事業として新CIシステムを導入。シンボルマークであるウイングマークは円の持つ調和と親しみ、正方形の持つ堅実と信頼のイメージを融和させ、暖かさ，確かさを表現し、斜めに走る3本の矩形は発展と活力を表象している。コーポレートカラーのフレッシュライトブルーは、確かさ，活力，親しみ易さ，新しさを表象している。
(五十嵐威暢)

142b

143a~143e
KAIGAI TSUSHO CI Plan
海外通商CI計画

Consulting
Communication Project Inc.
株式会社コミュニケーションプロジェクト

Art Director
Takenobu Igarashi
五十嵐威暢
Kazuhiro Hayase
早瀬和宏

Designer
Honami Morita
森田穂波
Yukimi Sasago
笹子行美

Client
KAIGAI TSUSHO K.K.
海外通商株式会社
UHAG (Switzerland)
ウハーグ(スイス)
1984

143a

In 1985 Kaigai Tsusho K.K. celebrated 60 years in import and sales of European high-precision machinery with the introduction of a new CI system. The company, which also provides integrated equipment systems, high-tech information and services, selected the motif to communicate the company's dedication to expansion and international exchange. The round exterior contour projects an image of harmony and intimacy, while the striped pattern evokes the ideas of attention to detail, precision, and advancement. This same logo is used for corporate identification purposes by UHAG, the company's head office in Switzerland, and all the COSA companies belonging to the worldwide UHAG group, promoting a uniform corporate image.
(Takenobu Igarashi)

1985年、ヨーロッパの高精密機械の輸入販売、総合的設備システムやハイテク情報及びサービスを提供する海外通商は、創業60周年を迎え新CIシステムを導入。コーポレートシンボルの円は調和と親しみを、ストライプは精度感、精密感、先進性を、2重の輪は国際的交流、広がりを表現している。同時にスイスの本社UHAGおよび全世界のグループ各社COSAも同シンボルによってCIを導入し、コーポレートイメージが統一された。
(五十嵐威暢)

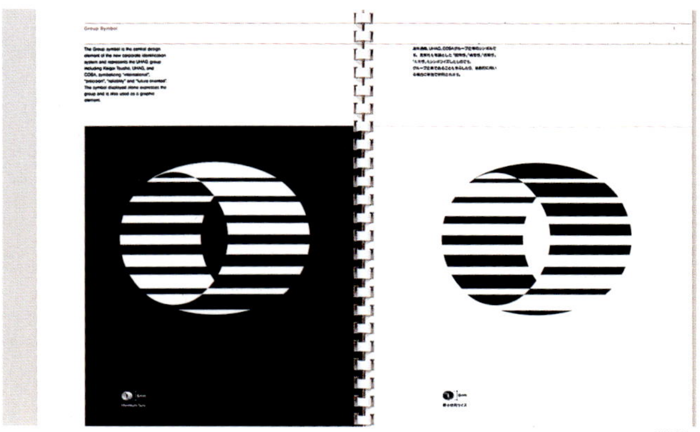

143b

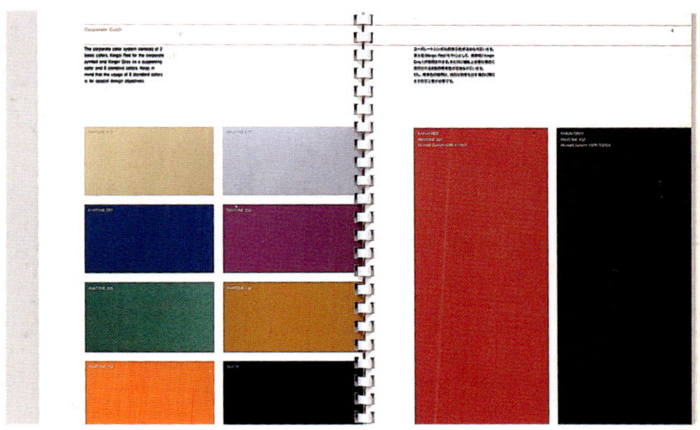

143c

143d

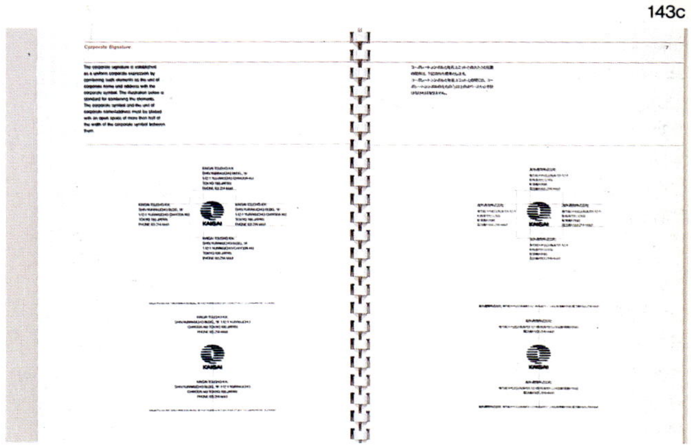

143e

144a〜144d
Meiji Milk Products CI Plan
明治乳業CI計画

Consulting
Dentsu Inc.
株式会社電通

Art Director & Designer
Takenobu Igarashi
五十嵐威暢

Designer
Yukimi Sasago
笹子行美
Kazuhiro Hayase
早瀬和宏

Client
Meiji Milk Products Co., Ltd.
明治乳業株式会社
1986

144a

In anticipation of the 1987 celebration of the 70th anniversary of its founding, Meiji Milk Products instituted a new CI system in 1986. The design of the initial "M" is intended to communicate the company's vitality and prospects for continued growth in the years leading up to the 21st century. It also expresses Meiji's corporate sturdiness and reliability. The six stripes represent the human body, life, culture, medicine, food, and health, standing for Meiji Milk Products roles and the areas of business in which it is active. The corporate color, red, signifies "verve," "warmth," "humanity."
(Takenobu Igarashi)

1987年創業70周年を迎える明治乳業は、1986年より新しくCIシステムを導入した。コーポレートシンボルのMEIJIは明治乳業の頭文字であるMに、躍動感と、21世紀に向かって伸びゆく明治乳業を象徴し、企業スケールと信頼感を表現した。6本のストライプは、からだ，生活，文化，医，食，健康という、明治乳業の役割と事業領域を意味している。コーポレートカラーのMEIJIレッドは情熱，暖かさ，人間愛を意味している。
(五十嵐威暢)

144b

144c

144d

145a~145b
NEOCON19 VI Plan
ネオコン19 VI計画

Art Director & Designer
Takenobu Igarashi
五十嵐威暢

Designer
Devi Shimamoto
デビ・島本

Client
The Merchandise Mart
ザ・マーチャンダイズマート
1986

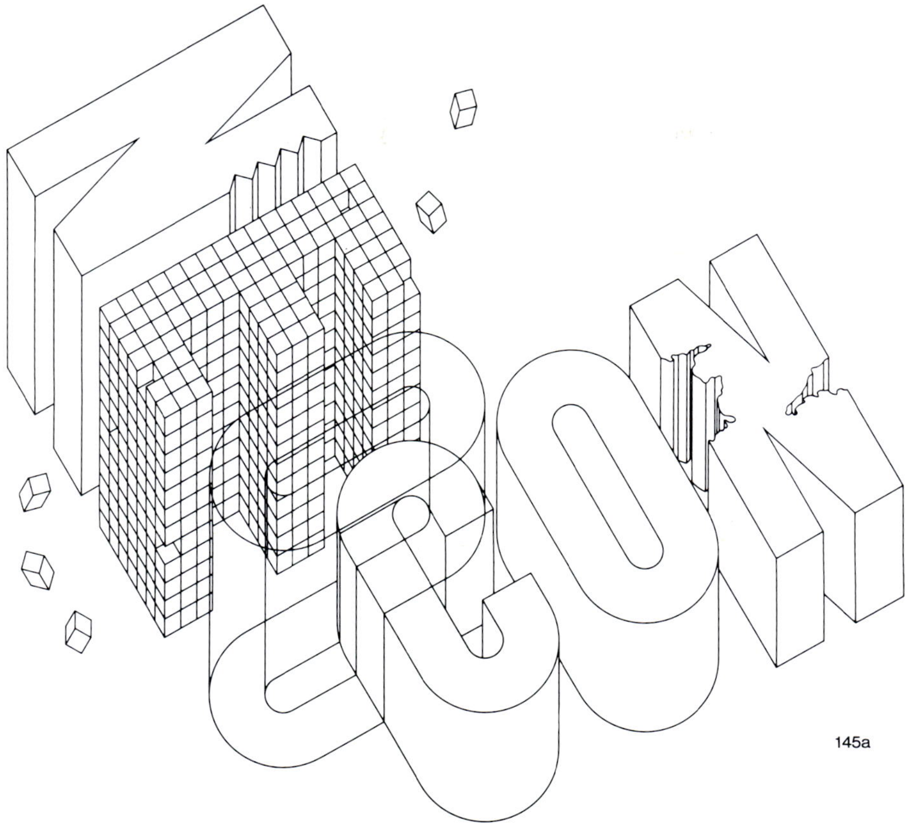

145a

145b

In 1986, 50,000 architects and designers from throughout the world will converge on Chicago for a 4-day trade fair-cum-convention that in the 19th in the NEOCON series, The Merchandise Mart largest annual event. The logo design for NEOCON 19 is both striking and immediately recognizable, as befits the identifying mark of a world-scale event in the field of interior and environmental design.
(Takenobu Igarashi)

シカゴのマーチャンダイズ・マートが行う年間最大行事のNEO-CONは今年で19年目となり、参加者も4日間に世界中から5万人の建築家とデザイナーが集中する。環境とインテリアに関する世界規模の会議と見本市に相応しく華麗で分かり易いイメージづくりを意図した。
(五十嵐威暢)

146a〜146c
Polygon Pictures CI Plan
ポリゴンピクチュアズCI計画

Art Director
Takenobu Igarashi
五十嵐威暢

Designer
Takenobu Igarashi
五十嵐威暢
Yukimi Sasago
笹子行美

Client
Polygon Pictures Inc.
株式会社ポリゴンピクチュアズ
1987

POLYGON PICTURES

146a

146b

146c

This logo, created by computer graphic techniques, is composed of a series of squares positioned three-dimensionally in a ring, like slides in a projector rack. This configuration, emblematic of dynamism and an international scope of activity, is an explicit statement of the corporation's character. (Takenobu Igarashi)

矩形によって映像のフレームを、それが連続して輪となることで、動きと国際的な活動範囲を象徴した。実際にもシンボルはコンピューターによって描かれ、この会社の性格を明快に表す結果となっている。
(五十嵐威暢)

147a~147c
China Moon VI Plan
チャイナムーンVI計画

Art Director
Takenobu Igarashi
五十嵐威暢
Kazuhiro Hayase
早瀬和宏

Designer
Honami Morita
森田穂波

Client
China Moon
チャイナムーン
1986

147a

147b

147c

In creating a whole new type of Chinese restaurant, a visual identity scheme encompassing all areas of the building plan down to such details as tableware and staff uniforms was utilized in the design of China Moon. The logo and color scheme transform the traditional image of a Chinese restaurant, introducing a romantic and fashionable atmosphere.
(Takenobu Igarashi)

新しいタイプの中華レストランをつくり上げるために、店舗計画の全本的計画をはじめ、ユニフォームやテーブルウェアに至るまでのVI計画を行った。シンボル、カラープラン等、従来の中華レストランのイメージを一掃し、ロマンチックな雰囲気とファッショナブルな感覚を取り入れた。
(五十嵐威暢)

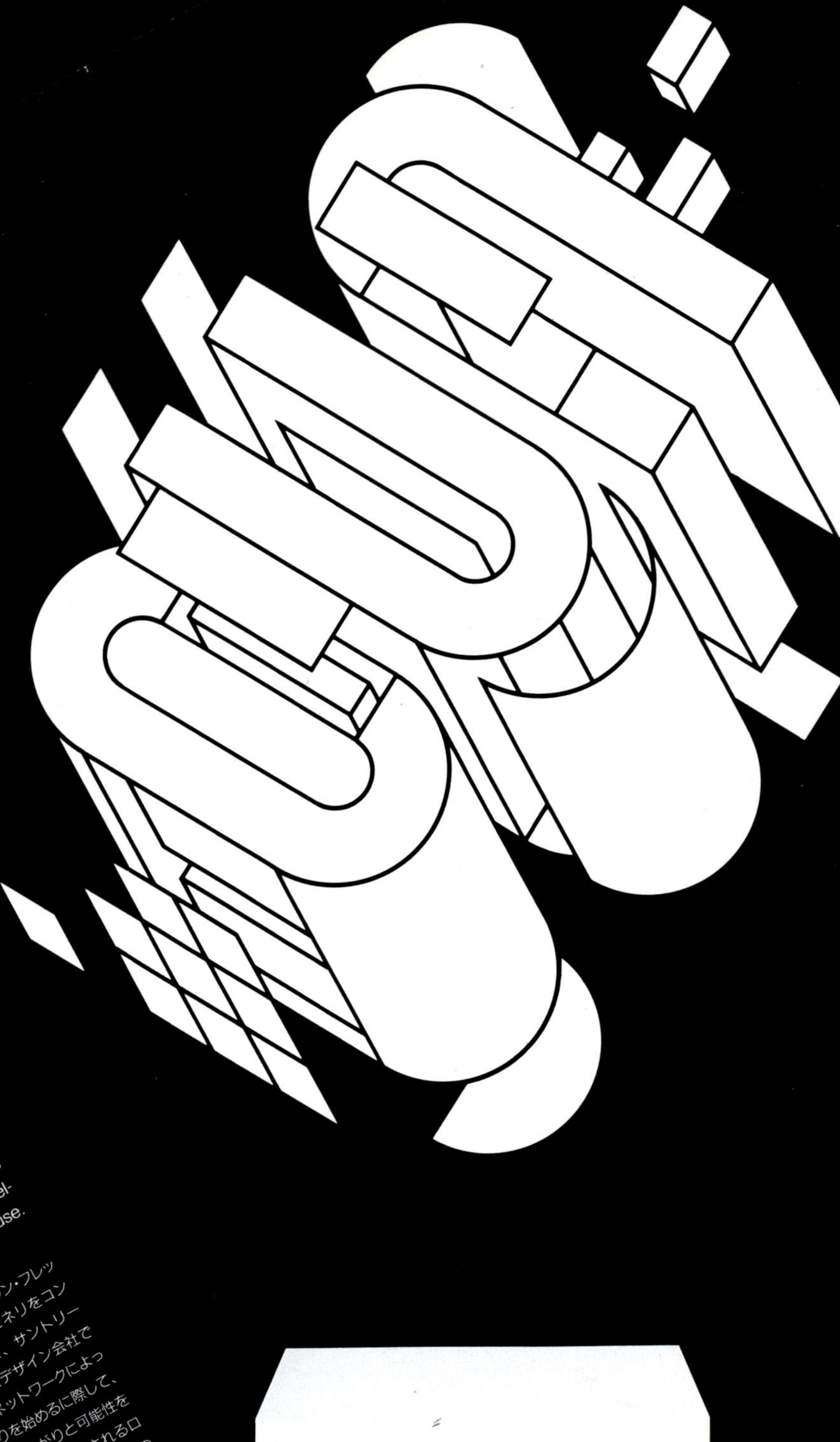

148a

...s
...ca-
...ypes
... devel-
...tual use.
...hi)

...暢、アラン・フレッ
...モ・ビネリをコン
...迎えて、サントリー
...立したデザイン会社で
...的なネットワークによっ
...い物作りを始めるに際して、
...Iも事業の広がりと可能性を
...現化して、永遠に増殖されるロ
...のデザインを持つ全く新しいCIの
概念を意図している。全てのアプ
リケーションは異なるロゴによっ
てデザインされ現在すでに50種類
が開発され使用されている。
(五十嵐威暢)

148b

Oun Corporation;Oun International Ltd.

148d

148e

148f

148g

148h

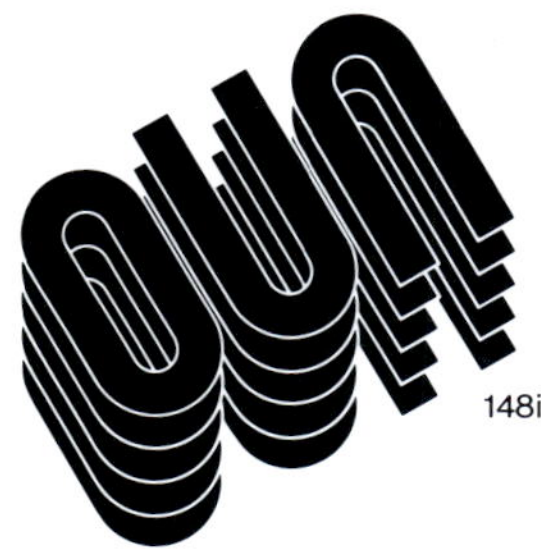
148i

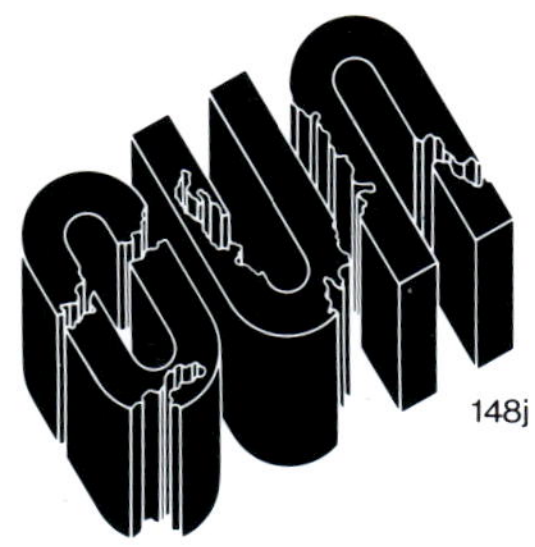
148j

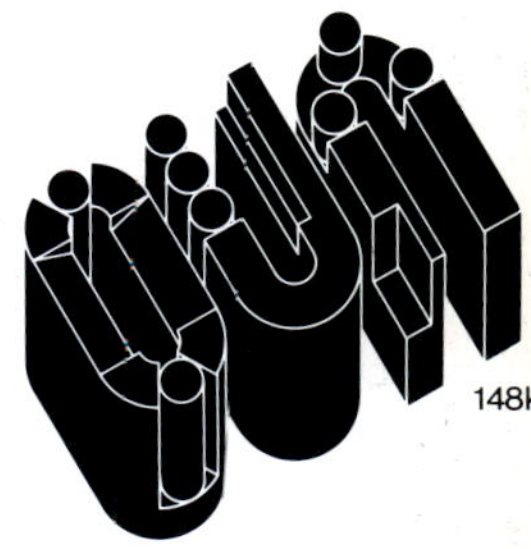
148k

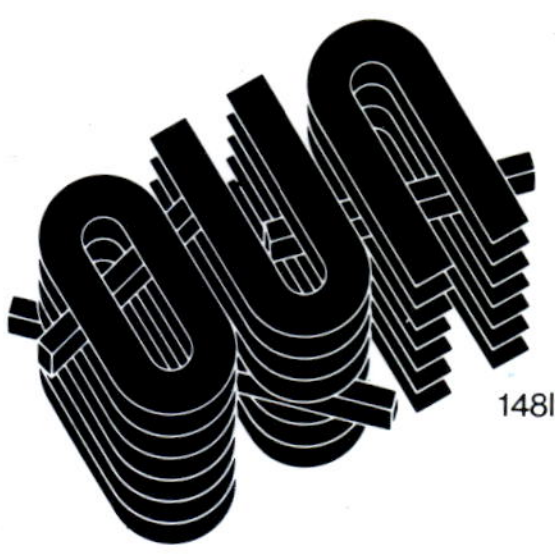
148l

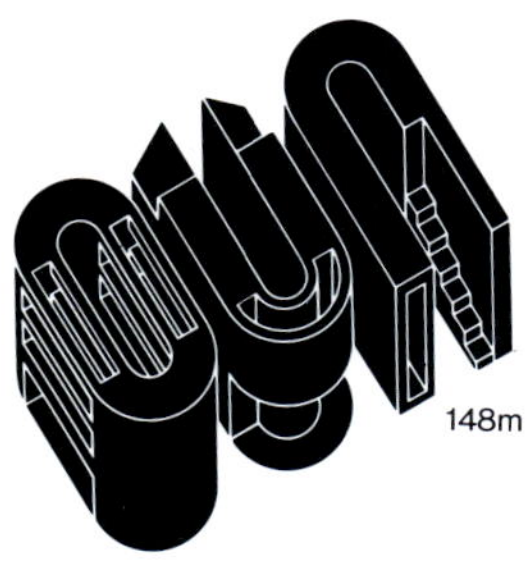
148m

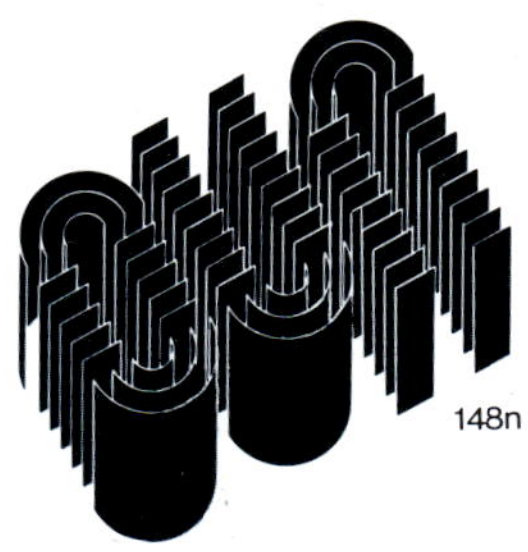
148n

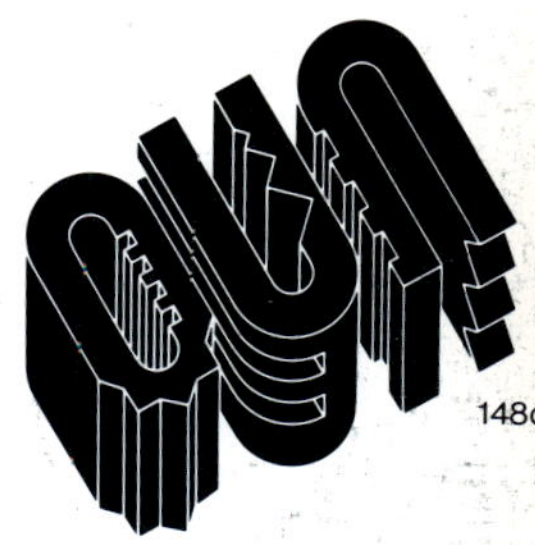
148o

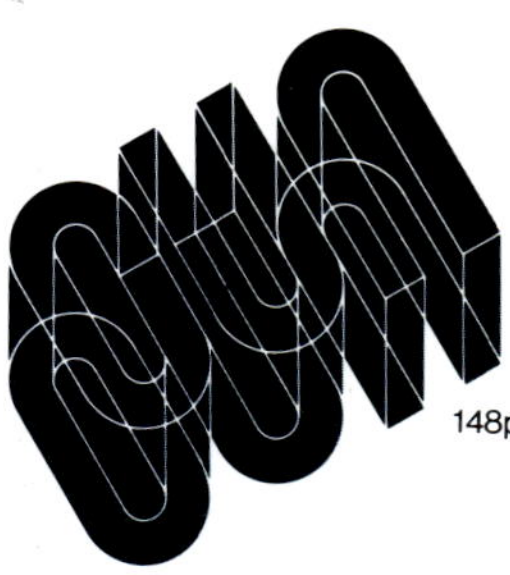
148p

148q

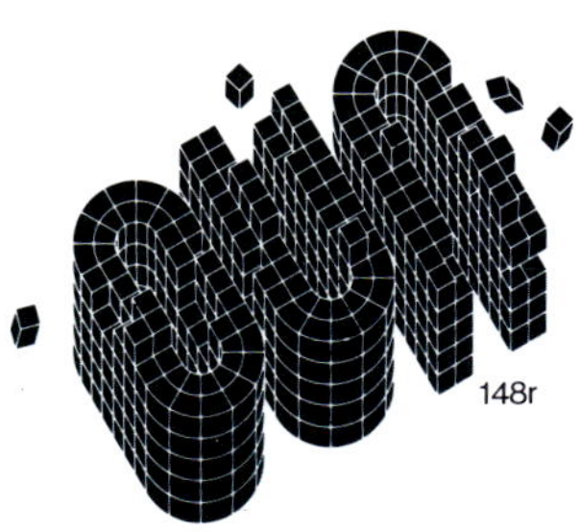
148r

148s

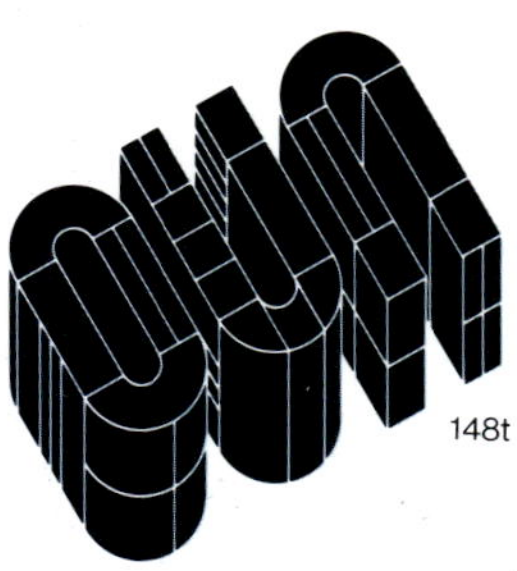
148t

148u

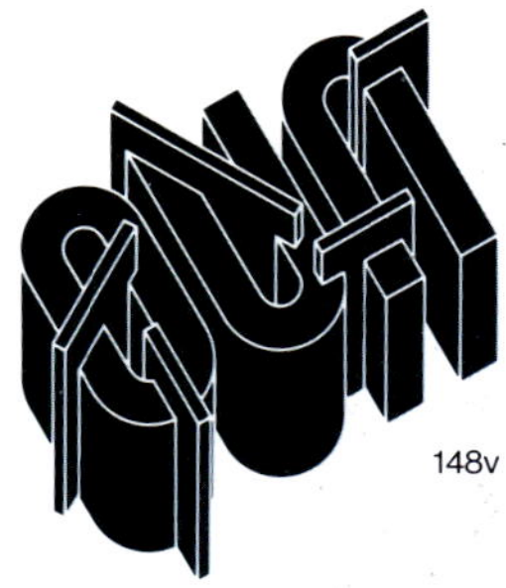
148v

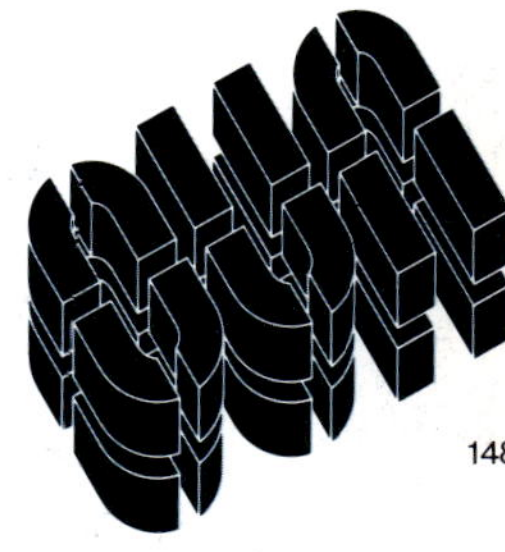
148w

149

Tokyo Prêt-à-porter Collection VI Plan

東京プレタポルテコレクションVI計画

Art Director & Designer
Takenobu Igarashi
五十嵐威暢

Designer
Yukimi Sasago
笹子行美

Client
Tokyo Prêt-à-porter Collection Office
東京プレタポルテコレクション事務局
The Yomiuri Shimbun
株式会社 読売新聞社
1985

Tokyo's first international prêt-à-porter collection was shown as an event to commemorate the 110th anniversary of the founding of the Yomiuri Shimbun. To identify Tokyo as a world center of fashion, I sought a bold logo design employing the two Chinese characters of the word "Tokyo."
(Takenobu Igarashi)

読売新聞創刊110周年を記念して、東京で初めての国際プレタポルテコレクションが開催された。ファッションの世界的発展地としての東京を位置づけるために、あえて漢字の東京二文字をモチーフとしてシンボルをデザインした。
(五十嵐威暢)

149

150

1st Winter Asian Games Symbol Mark

第1回冬季アジア大会 シンボルマーク

Consulting
Apt Corporation.
株式会社アプト
Dentsu Inc.
株式会社電通

Art Director
Takenobu Igarashi
五十嵐威暢

Designer
Yukimi Sasago
笹子行美

Client
Winter Asian Games Sapporo 1986 Office
冬季アジア大会事務局
1985

The winter-season event in the Asian Games series held every four years since the inaugural New Delhi Games in 1951 took place in the Japanese city of Sapporo for the first time in Japan in 1986. In the logo, an arrangement of white dots evokes simultaneously the images of a snow-covered mountain and of blowing snow. The oblique angle of the sets of dots also gives an impression of speed which governs the competition.
(Takenobu Igarashi)

1951年のニューデリー大会から4年に一度開催されているアジア大会の第一回冬季大会が1986年2月札幌で開催された。シンボルマークは白い点の集合によって降り続く雪と冬山のイメージを、左下に流れるレイアウトで競技のスピード感を表象している。
(五十嵐威暢)

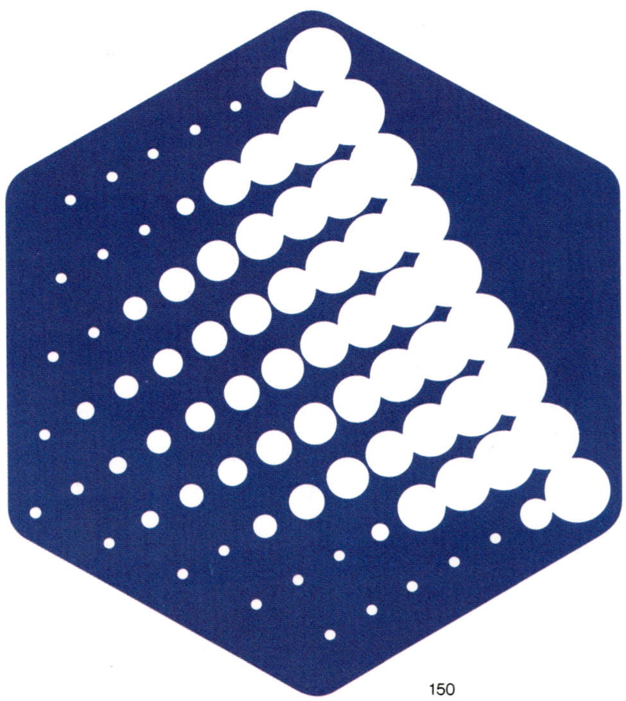

150

151
Kanazawa Institute of Technology UI Plan
金沢工業大学 UI計画

Art Director & Designer
Takenobu Igarashi
五十嵐威暢
Client
Kanazawa Institute of Technology
学校法人金沢工業大学
1983

This institute in Kanazawa, which has a unique software of its own and equipment for instruction in computer communications, is the first institution of higher education to introduce UI—university identity. The symbol for the university is based on the initials, K.I.T. with a golden eagle motif. It has been effectively used on the institute's stationery and for publicity materials and other name goods.
(Takenobu Igarashi)

最新のコンピューターコミュニケーション設備と独自のソフトウェアを有する金沢工業大学は、他の大学に先がけてUI(ユニバーシティ・アイデンティティ)を導入。シンボルマークは大学の略称であるK.I.T.とゴールデンイーグルをモチーフにデザインされ、ステーショナリー、広報物、各種グッズに効果的に展開されている。
(五十嵐威暢)

151

152
KISHO CORPORATION CI Plan
企商電子株式会社 CI計画

Art Director & Designer
Takenobu Igarashi
五十嵐威暢
Client
KISHO CORPORATION.
企商電子株式会社
1986

Kisho Denshi, a maker of audio products, introduced a CI scheme in 1986 to help establish its identity in the international market. This symbol is intended to express internationality, advancement, trust-worthiness and the idea of "energy in pursuit of an ideal."
(Takenobu Igarashi)

音響製品メーカーである企商電子は、国際マーケットにおける自社のアイデンティティの確立のため、1986年、CIを導入。シンボルマークは、国際性、先進性、信頼感と内から発散する"理想を追い求めるエネルギー"を表現している。
(五十嵐威暢)

152

153
Award for Leadership in Ending Hunger in Africa Symbol Mark
アフリカ飢餓救済リーダーシップアワード シンボルマーク
Art Director & Designer
Takenobu Igarashi
五十嵐威暢
Designer
Debi Shimamoto
デビ・島本
Client
The Hunger Project
ハンガープロジェクト
1987

The Africa Prize For Leadership
FOR THE SUSTAINABLE END OF HUNGER

153

154
GEM International Logotype
ジェムインターナショナル ロゴタイプ
Art Director
Takenobu Igarashi
五十嵐威暢
Designer
Yukimi Sasago
笹子行美
Client
GEM International Co., Ltd.
ジェムインターナショナル株式会社
1986

154

155
ACT-6 Logotype
アクト-6 ロゴタイプ
Art Director & Designer
Takenobu Igarashi
五十嵐威暢
Client
Tsuruya Department Co., Ltd.
株式会社鶴屋百貨店
1985

155

156
NOETIC・3D Logotype
ノエティック・3D ロゴタイプ

Art Director
Takenobu Igarashi
五十嵐威暢

Designer
Debi Shimamoto
デビ・島本

Client
Noetic Co., Ltd.
株式会社ノエティック
1985

156

157
APITA Seikatsu Soko Logotype
アピタ生活倉庫 ロゴタイプ

Art Director
Takenobu Igarashi
五十嵐威暢
Kazuhiro Hayase
早瀬和宏

Designer
Yukimi Sasago
笹子行美
Honami Morita
森田穂波

Client
Uni Co., Ltd.
ユニー株式会社
1985

157

158
PORTFOLIO Logotype
ポートフォリオ ロゴタイプ

Art Director
Takenobu Igarashi
五十嵐威暢

Designer
Yukimi Sasago
笹子行美

Client
Seibundo Shinkosha Publishing Co., Ltd.
株式会社誠文堂新光社
1985

158

159a〜159b

Alfa Records CI Plan

アルファレコードCI計画

Art Director & Designer
Eiko Ishioka
石岡瑛子

Designer
Motoko Naruse
成瀬始子

Client
Alfa Records, Inc.
アルファレコード株式会社
1980

ALFA

159a

Though its history is not so very long, the corporate mark and logo created at the time Alfa Records was founded soon ceased to suit the trend of the times and in 1980 it was forced to change to a whole new design. I took three points into consideration in creating this design: that it would not become outdated in 10 years' time; that it could be meaningful and appealing internationally as well as domestically; and that it could be reproduced in any kind of context and by any means.
(Eiko Ishioka)

歴史のさほど古くないレコード会社であるアルファレコードでも、設立当時のシンボルマークやロゴタイプのデザインが、時代の感性にあわなくなり、1980年には全く新しいデザインへと変更を余儀なくされ、私が、その新デザインを担当した。留意した3つのポイントは、次のようなものである。①デザインは10年先まで古くならないこと。②国際的に適用するレベルを持っていること。③どんな場でも、どんな手法でも再現可能なデザインであること。
(石岡瑛子)

159b

160a〜160b
THE GADD GANG BI Plan
ザ ガッド ギャングBI計画

Art Director & Designer
Eiko Ishioka
石岡瑛子

Designer
Keiko Motonaga
本永恵子

Photographer
Shinichiro Shishido
宍戸真一郎

Client
Epic/Sony Inc.
株式会社Epic/ソニー
1986

160a

"The Gadd Gang," the name of a New York-based jazz group led by drummer Steve Gadd, is also used as the title of their album. The logo design reflects the spirit and character of the band's music. Its principal motif is the initial "G.G." (Eiko Ishioka)

ニューヨークで活躍するジャズドラマーのスティーブ・ガッド率いるグループを、ガッドギャングと名付け、それをそのままアルバムのタイトルとした。彼等の創りだすキャラクターをロゴタイプとしてデザイン、イニシャルのG. G.が主たるモチーフとなった。(石岡瑛子)

160b

161a〜161b
A TOUCH VI Plan
ア タッチ VI計画

Art Director
Eiko Ishioka
石岡瑛子
Designer
Makoto Kumakura
熊倉 一
Photographer
Shinichiro Shishido
宍戸真一郎
Client
Epic/Sony Inc.
株式会社Epic/ソニー
1985

161a

Epic/Sony, a subsidiary of Japan's biggest record company, CBS/Sony, produces a wide range of recordings with a special emphasis on jazz. The "A Touch" label, targeting the global market, aims to present a new type of jazz that transcends the traditional boundaries of the genre.
(Eiko Ishioka)

日本最大のレコード会社 CBS/SONYの傍系会社 EPIC/SONYは幅広い音楽の中でも特にジャズに力を入れ、レコードを製作している。その中でも世界市場への進出を意図して登場させたレーベルが "A TOUCH" で、従来のジャズというジャンルを超えた新しいジャズミュージックづくりを目ざしたレーベルである。
(石岡瑛子)

161b

162

162
A Touch Of Love Logotype
ア タッチ オブ ラブ ロゴタイプ

Art Director & Designer
Eiko Ishioka
石岡瑛子
Designer
Makoto Kumakura
熊倉 一
Client
Epic/Sony Inc.
株式会社Epic/ソニー
1986

163a~163d
MILES DAVIS TUTU BI Plan
MILES DAVIS TUTU BIプラン

Art Director & Designer
Eiko Ishioka
石岡瑛子

Designer
Suzan Welt
スーザン・ウェルト

Photographer
Tsunehiro Kobayashi
小林庸浩

Client
Warner Bros. Records, Inc.
ワーナー・ブラザース
1986

miles davis TUTU

163a

Tutu is the first album put out by Miles Davis, the king of the jazz world, since he left CBS after over 30 years to sign with Warner Bros. Records, Inc. The album cover design combines a logotype based on the title with a portrait of Miles. Tutu received an award in the album package design division of the 29th Grammy Awards.
(Eiko Ishioka)

ジャズ界の帝王マイルス・デイヴィスが30年あまり在籍したCBSを離れ、ワーナー・ブラザーズに移って制作した初のアルバム。題名を素材としたロゴタイプとマイルスのポートレートを組み合わせてデザイン展開している。アルバム・パッケージ部門で第29回グラミー賞を受賞した。
(石岡瑛子)

163b

164a～164b
GOM BI Plan
GOM BI計画

Art Director & Designer
Eiko Ishioka
石岡瑛子
Designer
Motoko Naruse
成瀬始子
Client
Fuso Gom Industrial Co., Ltd.
富双ゴム工業株式会社
1984

The first works produced by architect Masayuki Kurokawa when he designed small interior accessories were made of gum (India rubber) and metal. The BI name "GOM" makes direct use of the Japanese word for these materials. The logo was designed to mirror the lines and mood of the pieces themselves.
(Eiko Ishioka)

建築家の黒川雅之が小物のデザインに取り組みはじめた初期の仕事は、ゴムと金属を結合させた商品である。GOMというBI名も、素材そのものの名称をダイレクトに使用している。デザインされた小物自体が持っている線と面の表情をそのまま生かし、ロゴタイプの個性とした。
(石岡瑛子)

165a～165b
ORIFA CI Plan
オリファ CI計画

Creative Director
Minoru Terada
寺田 稔
Art Director
Eiko Ishioka
石岡瑛子
Designer
Lance Wyman
ランス・ワイマン
Client
Orifa Co., Ltd.
株式会社オリファ
1984

Orifa is a personnel placement agency that aims to create a new paradigm in business. By excelling in quality/quantity of manpower and speed of dispatch, the company is helping to boost efficiency in business management. As extra-tall "A" with a steep incline was used in the logo to express the hope of its rapid growth, and it has in fact expanded at an eye-opening rate.
(Eiko Ishioka)

オリファはビジネスにおけるニューパラディアム(新しい規範)の創造をめざして、質、量、スピードの全面にわたる総合的なビジネスサービスの提供をし、企業経営の高効率化を手伝う人材派遣会社である。企業の急成長を願って、ORIFAのAに急勾配の直線を使用したところ実際に急成長をとげ、私を驚かせている。
(石岡瑛子)

gom
design: masayuki kurokawa
164a

ORIFA
Originative Industrial Faculty
165a

164b

165b

166a~166c
ISSEY MIYAKE PERMANENTE BI Plan
イッセイ ミヤケ ペルマネンテ BI計画

Art Director
Eiko Ishioka
石岡瑛子

Designer
Makoto Kumakura
熊倉 一

Client
ISSEY MIYAKE INTERNATIONAL Inc.
株式会社イッセイミヤケインターナショナル
1985

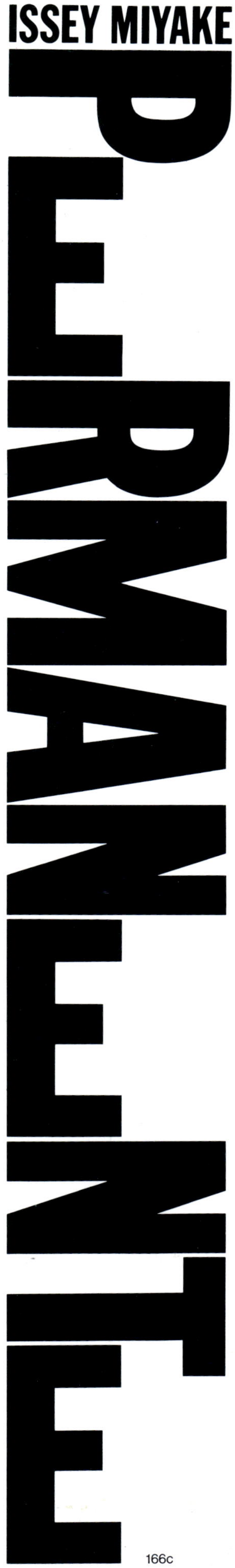

166c

166a

166b

In today's world of rapid shifts in fashion, it is common to lay aside superbly designed garments as dated after a single season. Questioning this practice, Issey Miyake has created a line of clothing under the brand name "Permanente." This Italian word translates as "permanent" in English. The brand name alone consists of ten letters.
(Eiko Ishioka)

すぐれたデザインの服がワンシーズンを経過しただけで古いデザインとして見捨てられていく傾向に疑問を持った三宅一生は永遠という名をつけたブランドで服を創りつづけている。イタリア語のPERMANENTEは英語のPERMANENT（永遠）を意味するが、なんと10にも及ぶアルファベットで成立っている。
（石岡瑛子）

167a~167b
KISSA CI Plan
KISSA CI計画

Art Director & Designer
Eiko Ishioka
石岡瑛子

Designer
Motoko Naruse
成瀬始子

Client
Kissa Co., Ltd
株式会社キサ
1977

KISSA

167a

167b

Kissa Takada, a pioneer in the field of shoe design, gave her own first name to both a company and a brand of shoes. The Kissa logo appears on packages, posters, catalogs, stationery, etc.
(Eiko Ishioka)

靴のデザイナーとしてパイオニア的存在の高田喜佐の会社名及びブランド名は彼女の名をとってKISSAとした。KISSAはパッケージ、ポスター、カタログ、ステーショナリー等に展開している。
(石岡瑛子)

168a〜168b
MISHIMA CI Plan
MISHIMA CI計画

Art Director
Eiko Ishioka
石岡瑛子

Calligrapher
Sharon Ann Nakazato
シャーロン・中里

Photographer
Shin chiro Shishido
宍戸真一郎

Client
Misima Production
ミシマプロダクション
1984

168a

168b

The American-made movie "Mishima" tells of the life and death of writer Yukio Mishima. It was the director's idea to have the title done in a calligraphic style. Since Mishima died by ritual self-disembowelment (hara-kiri), I tried to make the letters appear as if they were painted with splattered blood. But as some calligraphic works are difficult to read, I was careful to make the title legible.
(Eiko Ishioka)

アメリカ映画"MISHIMA"は三島由紀夫の生と死を題材にした映画である。タイトルを書で表現したいという監督の意図のもとに、切腹によって死をとげた三島由紀夫の血しぶきで描写したような文字を意図している。映画が世界配給であるため誰にでも読めることが重要なポイントとなった。
(石岡瑛子)

169a〜169d
INGOT BI Plan
INGOT BI計画

Art Director & Designer
Eiko Ishioka
石岡瑛子

Designer
Yumiko Iizuka
飯塚由美子
Masaya Maemura
前村昌也

Client
Nippon Light Metal Company, Ltd.
日本軽金属株式会社
1986

169a

169b

169c

169d

Three designers — Shiro Kuramata, Masayuki Kurokawa, and Toshiyuki Kita — formed a team to design interior goods made from aluminum. The name Ingot, used for BI purposes, signifies a mass of cast metal. Three different logotypes in all are employed to express the personality of each of the designers. The three can be used in any combination to create a new kind of BI image.
(Eiko Ishioka)

倉俣史朗、黒川雅之、喜多俊之の3人のデザイナーたちがアルミニウムを素材にインテリア・グッズをデザインした。INGOTというBI名は、金属の鋳魂を意味している。3人のデザイナーの個性を表現するため、三種類のロゴを起用。自由自在に組み合わせることで新しいBIのイメージ展開を試みている。
(石岡瑛子)

170a〜170b

The International Garden and Greenery EXPO '90 Osaka CI Plan

国際花と緑の博覧会CI計画

Symbol Mark Designer
Mitsuo Katsui
勝井三雄

Client
Japan Association for the International Garden and Greenery Exposition
財団法人 国際花と緑の博覧会協会
1987

170a

This design representing a single glowing bloom is symbolic of the mysteries of life, which endure in the face of technological advances. The "discs" that form the petals of the flower appear to have all turned to face the sun. But because no one disc is independent of the whole, we are reminded that a flower exists in a natural matrix of water and air—the indispensable elements of life itself. This flower logo stands as a paean to the harmony between the green earth and the delicate but hardy forms of life that inhaibt it.
(Mitsuo Katsui)

あまりにも急速な科学技術の進歩の中で、生命の神秘さを一輪の燁く華として象徴したものである。花びらを表わす円の変化の組み合わせは、すべて太陽の光に向かっての回転を示し、そのひとひらひとひらが、水・空気・温度といった生命そのものの不可欠な要素を結びつけ、共存している様子を表わしている。これは、緑の地球とそこに息づく優美で力強い小さな命が、共感する讃歌のシンボルとして作られたものである。
(勝井三雄)

170b

171a〜171b
Inter Design '85 Kyushu, Japan VI Plan
日本文化デザイン会議'85九州シンポジウムVI計画

Art Director & Symbol Mark Designer
Mitsuo Katsui
勝井三雄

Designer
Yutaka Sugimoto
杉本　豊

Client
Japan Inter Design Conference
日本文化デザイン会議事務局
1985

171a

This logo was created to integrate the visual schemes of the "Cultural Design" conferences held simultaneously in Kyushu's Kumamoto and Oita prefectures. Its triangular form approximates the geographic shape of the Kyushu land mass. Deep notches at the left side and top right represent the positions of Kumamoto and Oita and symbolize the theme of the symposium—"a cosmopolitan image." (Mitsuo Katsui)

九州の熊本と大分両県で同時に開催される文化デザイン会議のヴィジュアル計画のために制作されたものである。三角形によって九州全体を印象づけ、熊本と大分の位置である左右に細かいひだをつけることによって、コスモイメージをテーマにしたシンポジウムをシンボライズした。
(勝井三雄)

171b

172
Ajinomoto CI Plan
味の素CI計画

Art Director & Designer
Mitsuo Katsui
勝井三雄

Client
Ajinomoto Co., Ltd.
味の素株式会社
1969

Created as a commemorative symbol of the 60th anniversary of the company's founding in 1969, this logo has since been used in place of the old corporate emblem. Its simple, bold design, standing only for the "A" in Aji no Moto, has the universality to hold its own through changes in corporate personality.
(Mitsuo Katsui)

創業60周年の記念シンボルとして制作されたもので、旧社章にかわって今日まで使用されている。味の素のAという記号的な無性格なものをあえて選んだ。企業体質の変化に十分に耐えられるように考えられたものである。
（勝井三雄）

172

173
Yamada Shomei Symbol Mark
山田照明 シンボルマーク

Art Director & Designer
Mitsuo Katsui
勝井三雄

Designer
Yuji Dairokuno
大六野雄二

Client
Yamada Shomei Lighting Co., Ltd.
山田照明株式会社
1986

The change in Yamada Shomei's mark was made in the effort to do away with the outdated image of lighting as a simple, technically oriented industry. One of many companies catering to the latest in tastes and lifestyles, Yamada changed its logo from Japanese characters to Roman letters and adopted a design symbolizing the soft touch, brightness and friendliness.
(Mitsuo Katsui)

山田照明のロゴ変更は、照明という、単一業種で技術中心の古いイメージを消すことであった。ニューライフ創造関連会社として、新しい計画のもとに、ソフト、したしみやすさ、明快さを中心に、和文表記から"YAMADA"に変更、制作された。
（勝井三雄）

173

174
Roma Iwashima Symbol Mark
ローマ岩島 シンボルマーク

Art Director & Designer
Mitsuo Katsui
勝井三雄

Client
Roma Iwashima
ローマ岩島
1974

174

175
Lake Biwa Research Institution Symbol Mark
琵琶湖問題研究機構 シンボルマーク

Art Director & Designer
Mitsuo Katsui
勝井三雄

Client
Lake Biwa Research Institute
滋賀県琵琶湖研究所
1982

175

176
M. gallery Symbol Mark
エム ギャラリー シンボルマーク

Art Director
Mitsuo Katsui
勝井三雄

Designer
Yuji Dairokuno
大六野雄二

Client
M. gallery
株式会社 エム ギャラリー
1985

176

177
Kunishiro International Symbol Mark
国代耐火工業所 シンボルマーク

Art Director & Designer
Mitsuo Katsui
勝井三雄

Client
Kunishiro International, Inc.
株式会社国代耐火工業所
1976

177

178
Tokyo Shoseki Symbol Mark
東京書籍 シンボルマーク

Art Director
Mitsuo Katsui
勝井三雄

Designer
Yuji Dairokuno
大六野雄二

Client
Tokyo Shoseki Co., Ltd.
東京書籍株式会社
1986

178

179
Kodansha Video Logo-Mark
講談社ヴィデオ ロゴマーク

Art Director
Mitsuo Katsui
勝井三雄

Designer
Yuji Dairokuno
大六野雄二

Client
Kodansha Co., Ltd.
株式会社講談社
1986

179

180
Tosho Printing Symbol Mark
図書印刷 シンボルマーク

Art Director & Designer
Mitsuo Katsui
勝井三雄

Client
Tosho Printing Co., Ltd.
図書印刷株式会社
1967

180

181
CAST Logotype
キャスト ロゴタイプ

Art Director & Designer
Mitsuo Katsui
勝井三雄

Client
Kubota Ltd.
久保田鉄工株式会社
1968

181

182
The Yachiyo Shinkin Bank Symbol Mark
八千代信用金庫 シンボルマーク

Art Director
Mitsuo Katsui
勝井三雄

Designer
Yasuhiko Suzuki
鈴木康彦

Client
The Yachiyo Shinkin Bank
八千代信用金庫
1984

182

183a〜183g
CUISINE SHISEIDO BI Plan
キュイジーヌ・シセイドー BI計画

Art Director
Isamu Hanauchi
花内　勇

Designer
Masayoshi Nakajo
仲條正義
Akio Miyake
三宅章夫

Client
Shiseido Parlor Co., Ltd.
株式会社資生堂パーラー
1985

183a

I wanted to create a logo that evoked the new as well as the old. The design employs a checkerboard theme with the lettering worked into the graphic scheme. For the sake of variety, two colors—black and a rose that is close to a primary red—are used both as background and foreground colors.
(Masayoshi Nakajo)

新しいが、懐古的な面も出したかった。ロゴを市松のように組み合わせ、方形がそれを囲むというテーマである。色は原色に近いローズ色のポジとネガ、黒のポジとネガのヴァリエーションがある。
(仲條正義)

183b

183c

183d

183e

183f

183g

184a〜184d
WACOAL SPIRAL CI Plan
ワコール・スパイラルCI計画

Art Director
Shin Yoshida
吉田 臣
Designer
Masayoshi Nakajo
仲條正義
Client
WACOAL CORPORATION.
株式会社ワコール
1984

spiral

184a

5·6
minami
·aoyama
minato·ku
tokyo

184b

This simple name-based logo uses small letters only but can freely be combined with the names of the various booths within the building. The design follows a system of linkages between the letters. Use of color is unrestricted.
(Masayoshi Nakajo)

小文字のみで、単純なロゴであるが、ビル内の各ブースの名称との組み合わせが自由で、その組み合わせは、文字の一部（一点）で連結するというシステムを持っている。色は限定していない。
(仲條正義)

184c

184d

185a~185d
Matsuya CI Plan
松屋CI計画

Consulting
PAOS Inc.
株式会社パオス

Designer
Masayoshi Nakajo
仲條正義
PAOS Inc.
株式会社パオス

Client
Matsuya Co., Ltd.
株式会社松屋
1978

The client wanted a design totally different from logos reminiscent of family crests typical of the department stores that originated as kimono clothier shops or of the "friendliness"-stressing marks of the railway corporation stores. It needed to be cosmopolitan, refined, elegant, and, because of the store's location in the center of the ultra-sophisticated Ginza, targeted to the urban and urbane adult.
(Motoo Nakanishi)

This design was my entry in the competition among four designers for the new Matsuya CI concept done in 1984. It came into being quite naturally and I enjoyed the project immensely. The blue was originally a brighter, more striking hue, but shopping bags of that color looked soiled when dampened in the rain so it was later changed through the cooperation of Motoo Nakanishi of Paos and Matsuya.
(Masayoshi Nakajo)

従来の呉服店系百貨店の紋章風マークとも、電鉄系百貨店の親しみ強調のマークともまったく異なった、都会らしさ、繊細さ、華やかさに加えて、銀座という立地上、シティアダルト（都市型の大人）を主ターゲットとするデザインが意図された。4名の指名デザイナーによるコンペティションの結果、仲條正義氏のデザイン案が採用され、その後の多彩なデザイン展開、イメージ一新によって、業績低迷の百貨店が見事に甦っていった代表的なCI事例である。
（中西元男）

1978年の作品である。4人のデザイナーのコンペで、みな強敵であった。素直に仕上がり、楽しい仕事だった。採用されたペーパーバッグは、もっと明るく甘いブルーだったが、雨の舗道で濡れて汚れた色が不潔にみえたので、計画者のパオスの中西さんにお話しして、現在の色に変更をお願いした。一度社会に出たものを、私だけの希望で変えていただけたことはすばらしいことで、稀有なことだと思う。中西さんと松屋の雅量に感謝しなければならない。
（仲條正義）

MATSUYA
GINZA

185a

MATSUYA
GINZA
3-6-1 GINZA CHUO-KU TOKYO 〒104 ☎567-1211

185b

185c

185d

186
MG PLANET Logo-Mark
エム・ジープラネット ロゴマーク

Art Director
PAOS Inc.
株式会社パオス

Designer
Masayoshi Nakajo
仲條正義

Client
A Table Matsuya Inc.
株式会社アターブル松屋
1984

186

187a～187b
TACTICS DESIGN Logo-type
タクティクス デザイン ロゴタイプ

Art Director & Designer
Masayoshi Nakajo
仲條正義

Client
Shiseido Co., Ltd.
株式会社資生堂
1983

187a

187b

188

189a

189b

190

188
BISTRO BIS Logotype
ビストロ ビス ロゴタイプ

Art Director & Designer
Masayoshi Nakajo
仲條正義

Client
City Takahashi Co., Ltd.
株式会社シティ タカハシ
1984

189a〜189b
Hotel New Hakodate Logo -Mark
ホテルニューハコダテ ロゴマーク

Art Director & Designer
Masayoshi Nakajo
仲條正義

Client
Zenith Kigyo Co., Ltd.
ゼニス企業株式会社
1986

190
TERRA Logo-Mark
テラ ロゴマーク

Art Director & Designer
Masayoshi Nakajo
仲條正義

Client
Sapporo Furnishing Limited
札幌ファニシング株式会社
1986

191a〜191b

Shimoda PRINCE HOTEL VI Plan

下田プリンスホテルVI計画

Art Director & Symbol Mark Designer
Shigeo Katsuoka
勝岡重夫

Designer
Noboru Okata
大方　昇

Client
Izu Hakone Railway Co., Ltd.
伊豆箱根鉄道株式会社
1973

191a

This hotel needed a VI (Visual Identity) system that harmonized with the logotype system of the Prince Hotel chain yet reflected the expansive atmosphere of a resort. The logo developed to fulfill these requirements has as its motifs the blue ocean, white waves, and sea gulls of the Shimoda peninsula. It makes a strong statement of identity while evoking the pleasure and happy memories that accompany a visit to a resort. (Shigeo Katsuoka)

このホテルは、プリンス系ホテルのもつシンボルシステムから、さらにリゾートとしてのふくらみとイメージをもったVIシステムが要求された。これに応えるために開発されたのが、下田の青い海と白い波、カモメをモチーフとしたシンボルマークである。リゾートとしての楽しさ、想い出を演出するツールとして、このマークは強くアイデンティファイしている。
(勝岡重夫)

191b

192a〜192b
OFFSIDE BI Plan
オフサイドBI計画

Art Director & Logotype Designer
Shigeo Katsuoka
勝岡重夫

Designer
Takao Ikeda
池田隆男
Hiroko Tsukada
塚田宏子

Client
Kirin Brewery Co., Ltd.
キリンビール株式会社
1985

192a

As the name suggests, "Offside," a new Kirin beverage brand, does not fit the typical mold for a soft drink. Its unconventional taste and image are reflected in the logo and package design. Cooling beverages that target the youth market must be fresh and original in name and design.
(Shigeo Katsuoka)

キリンの新飲料・新ブランド戦略の「オフサイド」は、その名前通り従来の清涼飲料水のタブー、常識からの「ズレ・反則」感覚の商品であることをロゴタイプ、パッケージのデザインで強調した。ヤング向けの清涼飲料は、ネーミング、デザインに新鮮さ、奇抜さがあってよいし、又なくてはならない。
(勝岡重夫)

192b

193
YSS Logotype
YSS ロゴタイプ

Art Director
Shigeo Katsuoka
勝岡重夫

Designer
Toshio Hiraide
平出敏雄

Client
Nippon Gakki Co., Ltd.
日本楽器製造株式会社
1984

YSS
YAMAHA SPORTS STAFF

193

194
METAFA Logotype
メタファ ロゴタイプ

Art Director
Masanori Umeda
梅田正徳

Designer
Shigeo Katsuoka
勝岡重夫

Client
Arte Japan Ltd.
株式会社アルテジャパン
1983

NEO CONCEPTUAL SPACE
METAFA

194

195
Patchwork & Quilts Logo-type
パッチワーク&キルト ロゴタイプ

Designer
Shigeo Katsuoka
勝岡重夫

Client
Quilts Studio・Basket Ltd.
キルト工房・バスケット
1985

PATCHWORK
&QUILTS

195

196
Portable Keyboard Logo -Mark
ポータブル キーボード ロゴマーク

Art Director & Designer
Shigeo Katsuoka
勝岡重夫

Client
Nippon Gakki Co., Ltd.
日本楽器製造株式会社
1983

197
Toyo Glass Everyday Logotype
東洋ガラス Everyday ロゴタイプ

Art Director
Shigeo Katsuoka
勝岡重夫

Designer
Hideaki Shiota
塩田英明

Client
Toyo Glass Co., Ltd.
東洋ガラス株式会社
1986

198
INTER ROCK Logo-Mark
インターロック ロゴマーク

Art Director & Designer
Shigeo Katsuoka
勝岡重夫

Client
Inter Rock Corporation
インターロック株式会社
1986

199
MEDIART Logo-Mark
メディアート ロゴマーク

Consulting
Office 202 Co., Ltd.
株式会社オフィス 202

Art Director & Symbol Mark Designer
Shigeo Katsuoka
勝岡重夫

Logotype Designer
Hiroshi Asami
浅見 博
Takao Ikeda
池田隆男

Client
Mediart
株式会社メディアート
1983

MEDIART

199

200
NAF Symbol Mark
ナフ シンボルマーク

Art Director & Designer
Shigeo Katsuoka
勝岡重夫

Client
Naf, Ltd.
有限会社ナフ
1979

200

201
Nagai Symbol Mark
ナガイ シンボルマーク

Consulting
Office 202 Co., Ltd.
株式会社オフィス 202
Communication Project Inc.
株式会社コミュニケーションプロジェクト

Art Director & Designer
Shigeo Katsuoka
勝岡重夫

Client
Tokyo Nagai Co., Ltd.
株式会社東京ナガイ
1980

201

202
ITOX Symbol Mark
イトックス シンボルマーク

Consulting
Office 202 Co., Ltd.
株式会社オフィス 202

Art Director & Designer
Shigeo Katsuoka
勝岡重夫

Client
Itox
株式会社イトックス
1979

202

203
Doctor Homes Logo-Mark
ドクターホームズ ロゴマーク

Consulting
Dentsu Inc.
株式会社電通

Art Director
Shigeo Katsuoka
勝岡重夫

Designer
Hideaki Shiota
塩田英明

Client
Doctor Homes
ドクターホームズ株式会社
1986

203

204
HONDA Car Shop Symbol Mark
HONDA カー・ショップ シンボルマーク

Art Director & Designer
Shigeo Katsuoka
勝岡重夫

Client
Honda Autobox Co., Ltd
ホンダ オートボックス株式会社
1975

204

205
KAMEI Symbol Mark
カメイ シンボルマーク

Art Director & Designer
Shigeo Katsuoka
勝岡重夫

Logotype Designer
Hiroshi Asami
浅見 博

Client
Kamei Corporation
カメイ株式会社
1985

205

206
BLOX Logo-Mark
ブロックス ロゴマーク

Art Director
Shigeo Katsuoka
勝岡重夫

Logo-Mark Designer
Hiroshi Asami
浅見 博

Client
Maruichi Sailing Co., Ltd.
マルイチセーリング株式会社
1981

206

207
Star Lane Symbol Mark
スターレーン シンボルマーク

Art Director
Shigeo Katsuoka
勝岡重夫

Designer
Izumi Tanaka
田中 泉

Client
Star Lane Corporation
株式会社スターレーン
1983

207

208
New Project Symbol Mark
ニュープロジェクト シンボルマーク

Art Director & Designer
Shigeo Katsuoka
勝岡重夫

Client
New Project
有限会社ニュープロジェクト
1981

208

209
World Sports Fair Symbol Mark
国際スポーツフェア シンボルマーク

Art Director & Designer
Shigeo Katsuoka
勝岡重夫

Client
Fuji Television Network Inc.
株式会社フジテレビジョンネットワーク
1984

209

210
Meitetsu Department Store Symbol Mark
名鉄百貨店 シンボルマーク

Consulting
Dentsu Inc.
株式会社電通

Art Director & Designer
Shigeo Katsuoka
勝岡重夫

Designer
Hiroshi Asami
浅見 博
Takao Ikeda
池田隆男
Misako Inohara
猪原美佐子

Client
Meitetsu Department Store Co., Ltd.
株式会社名鉄百貨店
1984

210

211a〜211d

UV WHITE BI Plan

UVホワイトBI計画

Art Director
Yasui Kumai
熊井保亥

Logotype Designer
Helmut Schmid
ヘルムート・シュミット

Designer
Shuichi Ikeda
池田修一

Client
Shiseido Co., Ltd.
株式会社資生堂
1985

211a

Shiseido's UV White is a medicated cosmetic product designed to prevent skin damage from the effects of ultraviolet rays. The logo, UV, symbolizing this brand, is a graphic shorthand for the word "ultraviolet." In designing this logo, I aimed for simplicity, an elegance befitting a cosmetic product, and a strong statement of identity.
(Helmut Schmid)

UVホワイトは、資生堂の、紫外線によるトラブル予防の薬用化粧品である。ブランドシンボルUVは、ネーミングのウルトラ・ヴァイオレットを視覚化したものである。シンプル、洗練、コスメティックにふさわしいエレガンスを備えた、強いアイデンティティをねらってデザインした。
(ヘルムート・シュミット)

211b

211c

211d

212a〜212b

BECAUSE BI Plan

ビコーズBI計画

Art Director
Shunsaku Sugiura
杉浦俊作

Logotype Designer
Helmut Schmid
ヘルムート・シュミット

Designer
Shuichi Ikeda
池田修一

Client
Shiseido Co., Ltd.
株式会社資生堂
1984

BECAUSE

212a

212b

The toughest problem of a brand name beginning with the letter "B," as it does Shiseido's line of men's cosmetics, in creating a logo is to make it look like a "class A" logo. In the end the problem was easily solved. By design stress on the A in the middle, we came up with a striking, distinctive logo.
(Helmut Schmid)

資生堂男性化粧品ビコーズのロゴ制作にあたり、最大の難点は、Bで始まるロゴを、いかにしてAクラスに見せるか、ということだった。しかし、それは制作過程で簡単に解決した。中央に含まれるAを強調することで、記憶し易いA-ロゴができたのである。
(ヘルムート・シュミット)

213a〜213b
ELIXIR BI Plan
エリクシールBI計画

Art Director
Shunsaku Sugiura
杉浦俊作

Designer
Helmut Schmid
ヘルムート・シュミット
Shuichi Ikeda
池田修一

Client
Shiseido Co., Ltd.
株式会社資生堂
1983

ELIXR

213a

213b

The development of Elixir, a new cosmetic product by Shiseido, demanded a logo that symbolized its newness and the scientific basis for its claims of effectiveness. The clearly drawn lines of the lettering reflect the serious purpose of the product, and the point of intersection of the lines in the "x" is left open to symbolize the reversal of the aging process which is the effect of an elixir.
(Helmut Schmid)

資生堂の新・化粧品エリクシールの開発にあたり、新しさと化学性をシンボライズするロゴが要求された。クリアーな線のアルファベットでまじめさを表し、また、ネーミングのエリクシールは、Xの字に開放部分をつくり、エリクシールすなわちリバースをシンボライズすることを試みた。
(ヘルムート・シュミット)

214a〜214e
POCARI SWEAT BI Plan
ポカリスエットBI計画

Art Director & Designer
Helmut Schmid
ヘルムート・シュミット

Client
Otsuka Pharmaceutical Co., Ltd.
大塚製薬株式会社
1980

214a

One important aspect of the logo design for Pocari Sweat, Otsuka Pharmaceutical's ion-replenishing drink, is the two-tone wave-like figures which appear under the product name. These shapes represent "positive" and "negative" images of curves that illustrate the difference in the speed of moisture supply to the body before and after drinking Pocari Sweat. Combined with unevenly-spaced lettering in the product name, they create a compact, sharp, and workable logo that is easily remembered. (Helmut Schmid)

大塚製薬のイオンサプライ飲料ポカリスエットのデザインポイントは、ポカリ飲用前と後の、水分補給速度の違いを示すグラフ曲線を、ポジとネガのパターンで視覚化した点である。この曲線と、字間のアキの異なる商品名を組み合わせて、応用性に富んだ、コンパクトで記憶し易い、爽やかなシンボル・ロゴを狙った。
(ヘルムート・シュミット)

214b

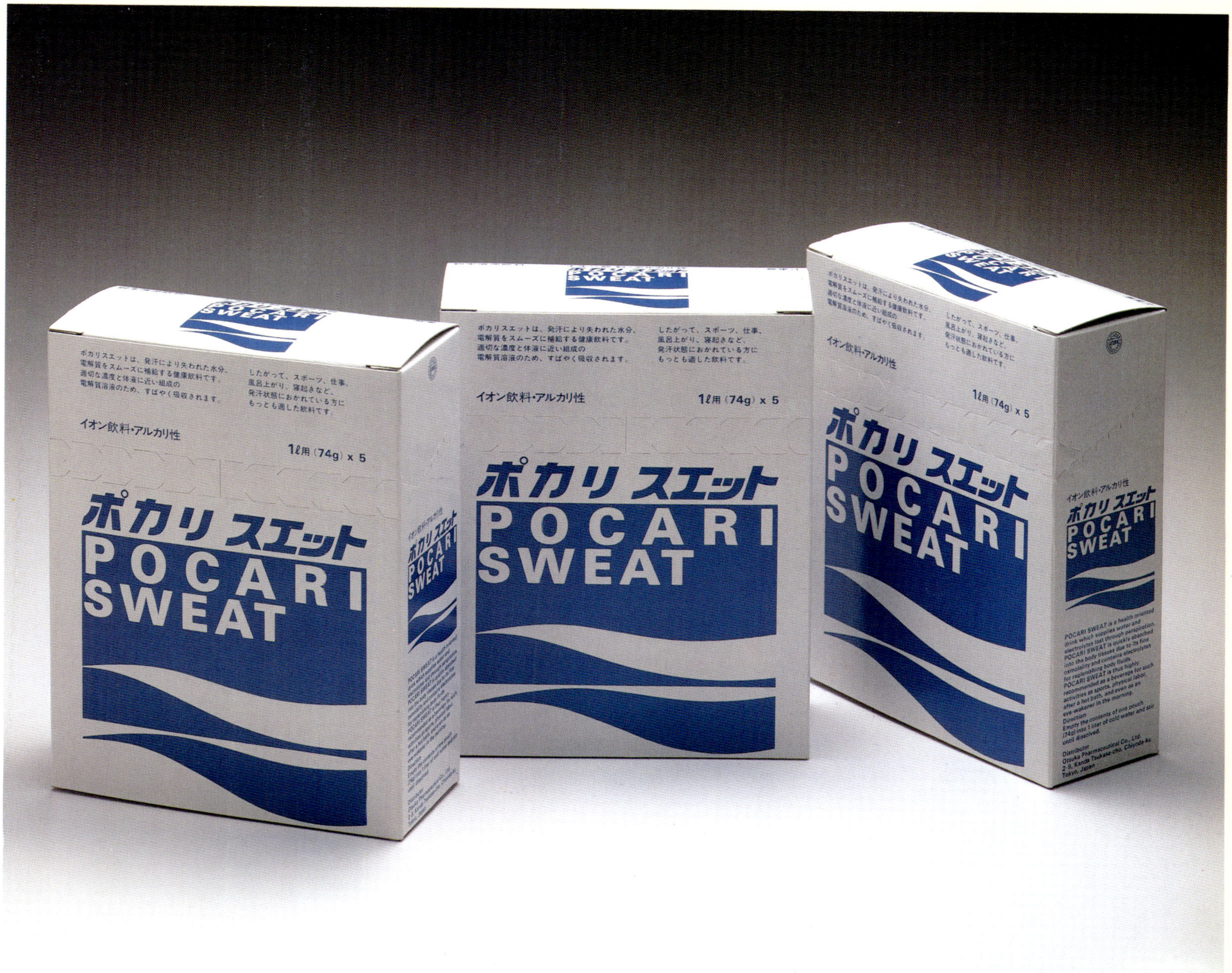

214c

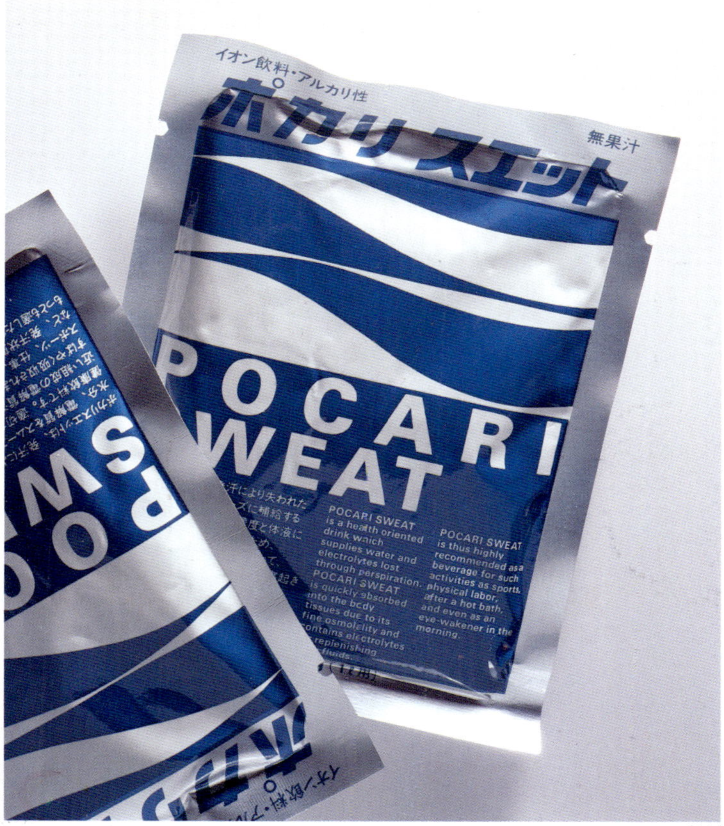

214d

214e

215a～215e
Kobe Belle CI Plan
神戸ベルCI計画

Art Director & Designer
Helmut Schmid
ヘルムート・シュミット

Client
Belle Co., Ltd.
株式会社ベル
1985

Belle is a Kobe-based confectionery with approximately 40 stores in Tokyo and the Kansai region. A new CI scheme was introduced this year, 1987, to mark the company's 60th anniversary, and the word "Kobe" was added to the corporate name to give it more impact and stress the fact that the company has its roots in Kobe. The new Kobe Belle logo is simple and charming, and these characteristics are underscored when it is used in a repetitive arrangement.
(Helmut Schmid)

ベルは神戸に本拠を置き、関西・東京に約40店舗を持つ洋菓子メーカーである。60周年記念として新しいCIが計画された時、ネーミングに神戸を加え、より印象強く、また神戸生まれという点を強調することを提案。新しい神戸のベルのロゴマークは、やさしさ・かわいさを表現し、特に連続して使う時にその特長が強調される。
(ヘルムート・シュミット)

215a

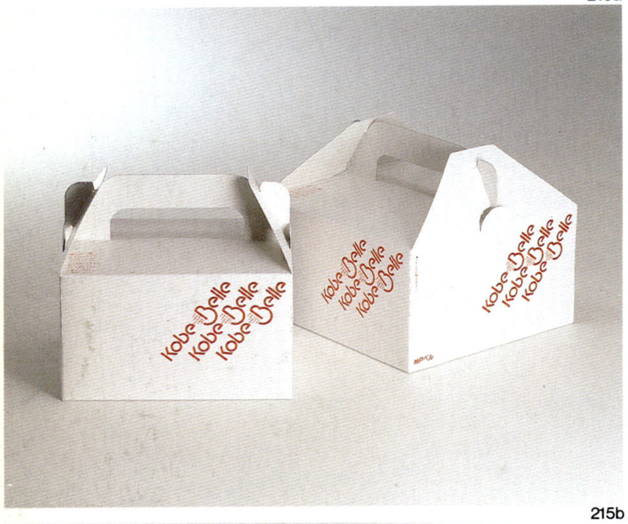

215b

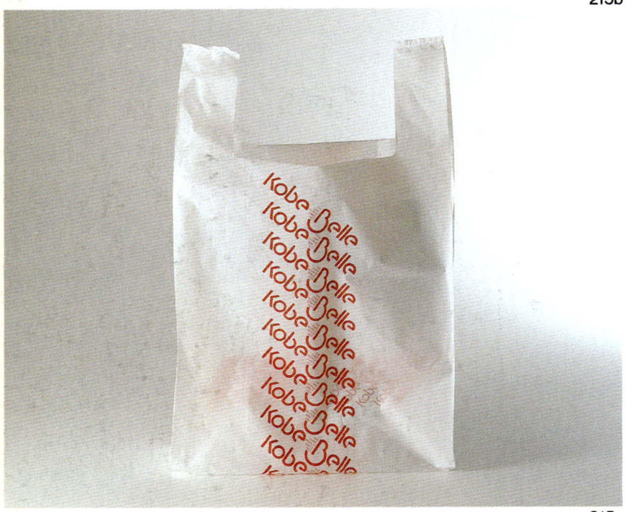

215c

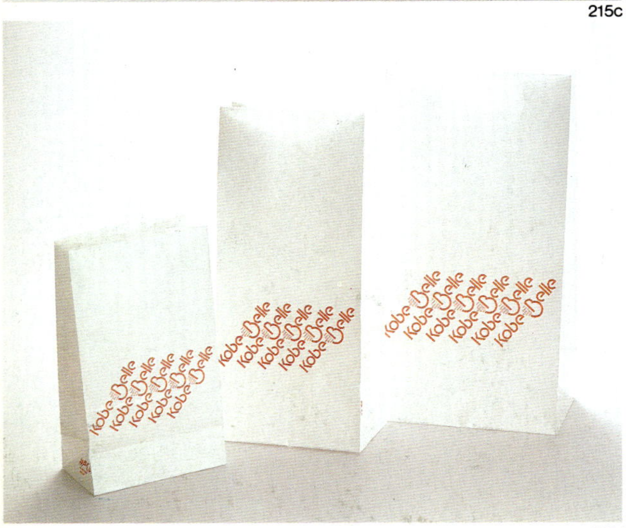

215d

215e

216a〜216e

RUBAN D'OR BI Plan

リュバンドールBI計画

Art Director & Designer
Helmut Schmid
ヘルムート・シュミット

Client
Belle Co., Ltd.
株式会社ベル
1981

216a

216b

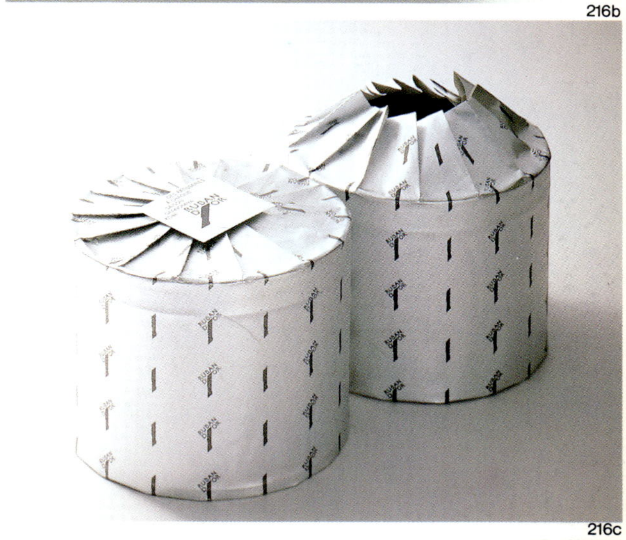

216c

216d

216e

Ruban D'or is a high-class confectionery with its corporate headquarters located in Kobe. The company's CI scheme centers on a logo that visually represents the corporate name, taken from a French expression meaning "gold ribbon." The logo also functions as a guarantee of quality and is used on packages as a kind of seal—not as a simple background design.
(Helmut Schmid)

リュバンドールは神戸に本拠を置く高級洋菓子のブティックである。CIは、金のリボンを意味する仏語のネーミングを視覚化したシンボル・ロゴを中心に展開した。このシンボル・ロゴには、品質証シール的役割も持たせ、それに適した使い方をし、決して地模様的扱いはしないこととした。
(ヘルムート・シュミット)

217a〜217d
SKOAL BI Plan
SKOAL BI計画

Art Director
Takahisa Kamijo
上條喬久

Designer
Eiji Hosokawa
細川栄二

Client
Kashiyama Co., Ltd.
樫山株式会社
1982

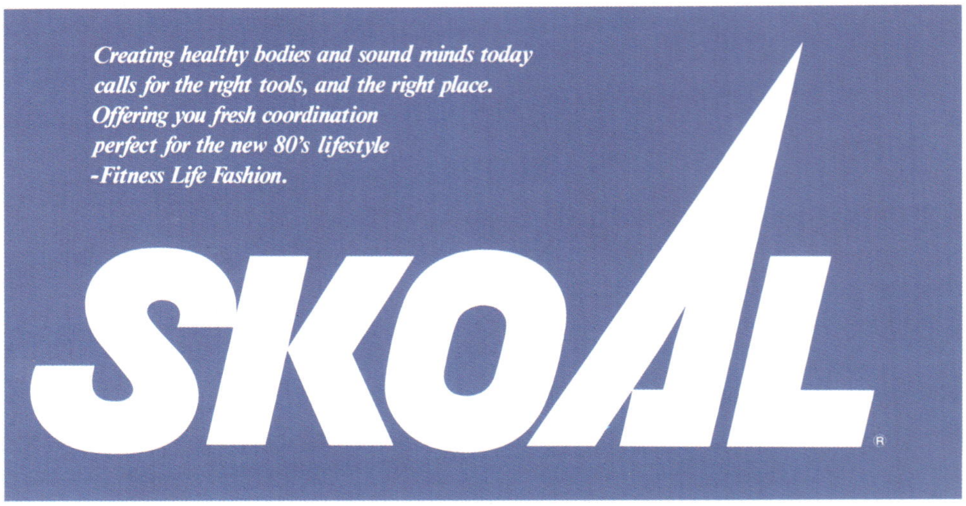

217a

The Skoal brand was created as Onward's line of genuine casual wear fashions, with an image of simplicity, smartness and vitality. For the logo I simply tried to give visual expression to the meaning and sound of the word Skoal. Visual expression and meaning came together quite well in this design. The layout of the logo is made generous in size in order to evoke the feeling of ample space.
(Takahisa Kamijo)

オンワードの本格的なカジュアルウェアとして、開発されたブランドであり、シンプルで爽やかなイメージと元気の良さが要求された。そこで、SKOALの意味と発音を素直に表現してみることにした。視覚表現と言葉の意味が、不思議なほど一致した。また、豊かな空間を感じさせるように、ロゴマークがレイアウトされる空間を広くとる演出も試みた。
(上條喬久)

217b

217c

217d

218a〜218f
FOR LIFE RECORDS, CI Plan
フォーライフCI計画

Art Director
Takahisa Kamijo
上條喬久

Designer
Eiji Hosokawa
細川栄二

Client
FOR LIFE RECORDS, INC.
株式会社フォーライフ レコード
1986

218a

218b

Ten years after FOR LIFE RECORDS was begun, FOR LIFE WING was born to take on newer challenges. A new image was also adopted for the established FOR LIFE brand. So a major theme in this CI scheme was the coordination and differentiation of these two labels. We sought coordination by retaining the old corporate colors and differentiation by forms that suggest a sense of musical rhythm.
(Takahisa Kamijo)

FOR LIFE RECORDS発足10年を機に、より新しい挑戦の目的で、新レーベルFOR LIFE WINGが誕生。同時に、従来のFOR LIFEレーベルもイメージを一新することになった。そこで、2つのレーベルの統一と変化がこのCIの重要なテーマとなった。従来のコーポレートカラーを受け継ぐことで統一を、音楽のリズム感をそれぞれの形にすることで変化を表現した。
(上條喬久)

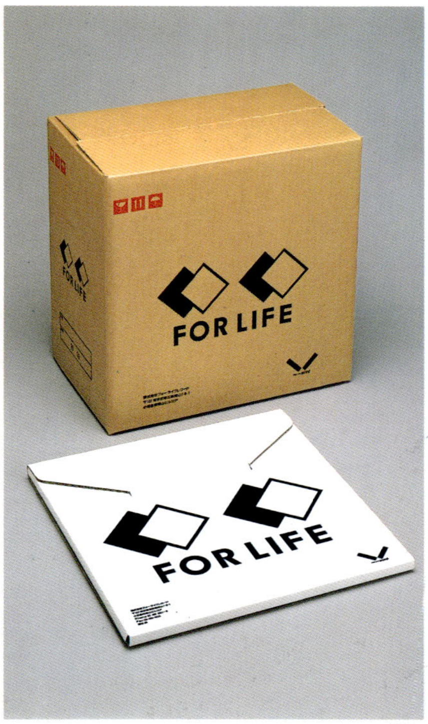

218c

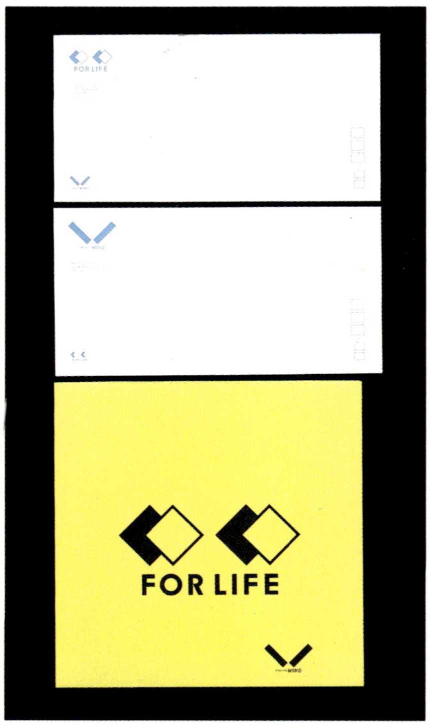

218d

218e

218f

219a〜219d
YONEX CI Plan
ヨネックスCI計画

Art Director & Designer
Takahisa Kamijo
上條喬久

Client
YONEX CO., LTD.
ヨネックス株式会社
1975

219a

219b

219c

219d

This new logo was created when Yoneyama Racket changed its name to Yonex Co., Ltd. Its corporate mark, redesigned with simpler lines, and the company colors—sky blue and forest green—were retained in the New CI scheme, so the adoption of Yonex's new corporate image went very smoothly.
(Takahisa Kamijo)

ヨネヤマラケットからヨネックスに社名変更された時点で、ロゴタイプを新しくした。マークは、よりシンプルに、リデザインして踏襲することになった。また、大空の青、大地の緑であるブルー&グリーンのコーポレートカラーも、そのまま新しいCIに採用したので、YONEXの新しいCIは、非常にスムーズにイメージの継続ができた。
(上條喬久)

220
The Japan Tobacco & Salt Public Corporation Symbol Mark
日本専売公社 シンボルマーク

Art Director & Designer
Takahisa Kamijo
上條喬久

Client
The Japan Tobacco & Salt Public Corporation
日本専売公社
1981

日本専売公社 220

221
LUFT Logo-Mark
ルフト ロゴマーク

Art Director & Designer
Takahisa Kamijo
上條喬久

Client
GROUP LUFT
グループ ルフト
1983

221

222
SOSS Logotype
エスオーエスエス ロゴタイプ

Art Director
Takahisa Kamijo
上條喬久

Designer
Eiji Hosokawa
細川栄二

Client
Sun Shine City Corporation
株式会社サンシャイン シティ
1986

222
SUNSHINE OPERATION SERVICE SYSTEM

223
HACHIYA Logo-Mark
蜂屋 ロゴマーク

Art Director
Takahisa Kamijo
上條喬久

Designer
Chiaki Nawa
名和千明

Client
Hachiya Co., Ltd.
蜂屋株式会社
1986

224
Tiens! Logo-Mark
ティアン ロゴマーク

Art Director & Designer
Takahisa Kamijo
上條喬久

Designer
Eiji Hosokawa
細川栄二

Client
Mimatsu Co., Ltd.
株式会社三松
1986

225
Taiyo Printing Logo-Mark
大洋印刷 ロゴマーク

Art Director & Designer
Takahisa Kamijo
上條喬久

Client
Taiyo Printing Co., Ltd.
大洋印刷株式会社
1985

HACHIYA 223

Tiens! 224

TAIYO PRINTING CO., LTD. 225

226
Big Process Logotype
ビッグプロセスロゴタイプ
Art Director & Designer
Takahisa Kamijo
上條喬久
Client
Big Process Co., Ltd.
株式会社ビッグプロセス
1975

227
TYR Logo-Mark
ティール ロゴマーク
Art Director
Takahisa Kamijo
上條喬久
Designer
Eiji Hosokawa
細川栄二
Client
Mimatsu Co., Ltd.
株式会社三松
1986

228
FORTUNE Logo-Mark
フォーチュン ロゴマーク
Art Director
Takahisa Kamijo
上條喬久
Designer
Eiji Hosokawa
細川栄二
Client
J・Osawa & Co., Ltd.
株式会社大沢商会
1983

229
Joséhine Logo-Mark
ジョゼフィーヌ ロゴマーク
Art Director & Designer
Takahisa Kamijo
上條喬久
Client
Mimatsu Co., Ltd.
株式会社三松
1984

230a～230d
The Hanshin Department Store CI Plan
阪神百貨店CI計画

Art Director & Designer
Jun Kusakari
草刈 順
Client
The Hanshin Department Store, Ltd.
株式会社阪神百貨店
1978

230a

230b

230c

230d

The Han-shin Department Store logo embellishes a graphic coalition of the initials "H" and "S" with a 4-leafed clover. The 4-leafed clover stands for happiness and hope, and the store's message to its clientele is the three simple words "Live in clover."
(Jun Kusakari)

阪神百貨店のイニシャル "H" と "S"を組み合わせたものに、四つ葉のクローバーをあしらった。四つ葉のクローバーは幸福あるいは希望を意味し、百貨店から生活者へのメッセージには "LIVE IN CLOVER"という言葉を使用している。
(草刈 順)

231a〜231b
PARCO CI Plan
パルコCI計画

Art Director & Designer
Jun Kusakari
草刈 順

Client
Parco Co., Ltd.
株式会社パルコ
1970

PARCO

231a

231b

Parco is a complex of stores specializing in clothing and household furnishings that aims to attract young female shoppers. Its logo design employs fashionable lettering with a ladylike elegance. No restrictions regarding color or application govern use of the logo. This gives it the flexibility to freely adapt to and reflect the changing times.
(Jun Kusakari)

パルコは若い女性を対象にしたファッションとリビングの専門店の集積体。女性らしい優雅さとファッショナブルなレタリングを心がけた。色や使用法には特に規定をつくらず、時代の流れに応じて自由な表現ができるような方針をとっている。
(草刈 順)

232a～232f
The Seibu Department Stores CI Plan
西武百貨店CI計画

Art Director & Designer
Jun Kusakari
草刈 順

Client
The Seibu Department Stores, Ltd.
株式会社西武百貨店
1975

232a

232b

Seibu's logo and mark of corporate identity are based not on abstract shapes but on the letters of the corporate name itself. For newspaper and other advertising purposes, a mark combining the Chinese characters for "Seibu" with the usual Roman lettering is often used.
(Jun Kusakari)

コーポレートマークには抽象的な形態をさけ、"SEIBU" というローマ字を主体にしたマーク・ロゴとした。別に、新聞広告などには「西武」という日本字と漢字を組み合わせた広告用の表示も設けてある。
(草刈 順)

232c

232d

232e

232f

233
Ikebukuro Shopping Park Symbol Mark
池袋ショッピングパーク シンボルマーク

Art Director & Designer
Jun Kusakari
草刈 順

Client
Ikebukuro Shopping Park Co., Ltd.
株式会社池袋ショッピングパーク
1981

233

234
Gour Cuss Logotype
グルカス ロゴタイプ

Art Director & Designer
Jun Kusakari
草刈 順

Client
Nankai Electric Railway Co., Ltd.
南海電気鉄道株式会社
1986

234

235a〜235c
WAGAMI BI Plan
WAGAMI BI計画

Art Director & Designer
Hiroshi Morishima
森島　紘
Client
Time-Space-Art Inc.
タイム・スペース・アート株式会社
1980

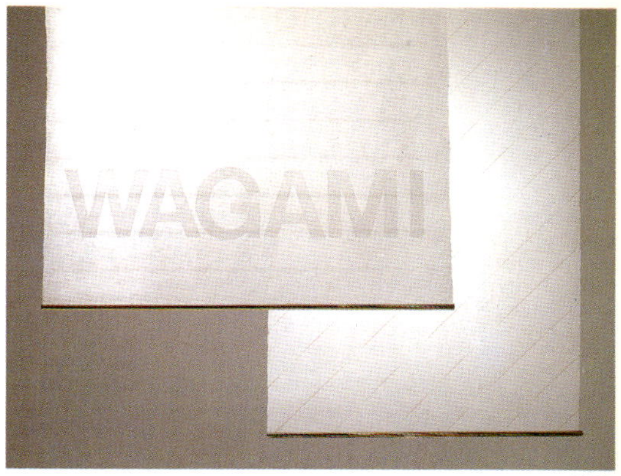

235b

WAGAMI
DESIGNS
TIME=SPACE=ART INC.

235a

235c

"Wagami"is a line of products made from Japanese paper. Its logotype, based on the helvetica and family type styles, hints at the international flavor of these products by using Roman characters to spell out the Japanese word "wagami." (Hiroshi Morishima)

和紙を素材にしたプロダクト製品。インターナショナルな感覚でのモノづくりを目指して"WAGAMI" とローマ字で標記。ヘルベチカ、ファミリーをベースにロゴタイプ化した。
(森島　紘)

236a〜236b
Taiyo Kobe Bank CI Plan
太陽神戸銀行CI計画

Art Director & Designer
Tadashi Ohashi
大橋　正

Client
Taiyo Kobe Bank Ltd.
株式会社太陽神戸銀行
1973

236a

This mark was created when the Taiyo Bank and Kobe Bank merged. The circle symbolizes the sun (taiyo) and the wave-like symbol in the center evokes the image of Kobe's famous international port. The uninterrupted loop form of the wave symbol is intended to express the idea of eternal growth and development. (Tadashi Ohashi)

太陽銀行と神戸銀行の合併によりつくられたマークである。円は太陽を、中央の波形は港・神戸を表現し、波形を無限記号にすることによって永遠の発展を意味させた。(大橋　正)

236b

237
Meiji Centennial Symbol Mark
明治百年 シンボルマーク

Art Director & Designer
Tadashi Ohashi
大橋　正

Client
Prime Minister's Office
総理府
1967

237

238
The Kyowa Bank Symbol Mark
協和銀行シンボルマーク

Art Director & Designer
Tadashi Ohashi
大橋　正

Client
The Kyowa Bank Ltd.
株式会社協和銀行
1968

238

239a〜239b
International Design Festival VI Plan
国際デザイン・フェスティバルVI計画

Art Director & Designer
Takeshi Otaka
大高　猛
Client
Japan Design Foundation
財団法人国際デザイン交流協会
1982

INTERNATIONAL
DESIGN
FESTIVAL,
OSAKA

239a

239b

The circle in this logo stands for the earth floating in space, Japan (land of the rising sun), and the "O" in Osaka while also representing the root of all design. The whirlpool pattern underneath—an old Japanese symbol for eddying water—is emblematic of life and the swirling activity of a festival, while resembling the letter "d" in "design." Through the union of these two shapes—still and dynamic—the logo creates a fitting symbol for festival projecting an image of the future. (Takeshi Otaka)

円は宇宙に浮かぶ地球、日本、そして大阪の「O」を意味し、同時にあらゆるデザインの原点をも意味しています。また下の巴は涌きたつ水を表わす日本古来のかたちであり、それは生命を具現し、デザインの「d」と、祭（フェスティバル）の姿を意味しています。そしてこの静と動の二つのかたちの結合によって、未来像を指向する国際デザイン・フェスティバルを表現しています。
（大高　猛）

240

241

242

243

240
Osaka Convention Bureau Symbol Mark
大阪コンベンション・ビューローシンボルマーク
Art Director & Designer
Takeshi Otaka
大高 猛
Client
Osaka Convention Bureau
大阪コンベンション・ビューロー
1984

241
Toyosu Symbol Mark
とよす シンボルマーク
Art Director & Designer
Takeshi Otaka
大高 猛
Client
Toyosu Co., Ltd.
株式会社とよす
1978

242
Mode U Symbol Mark
モード・ユー シンボルマーク
Art Director & Designer
Takeshi Otaka
大高 猛
Client
Mode U Co., Ltd.
モード・ユー株式会社
1987

243
Terauchi Group Symbol Mark
寺内グループ シンボルマーク
Art Director & Designer
Takeshi Otaka
大高 猛
Client
Terauchi Co., Ltd.
寺内株式会社
1982

244a～244b
SUNTORY Foods BI Plan
サントリーフーズBI計画

Art Director & Symbol Mark Designer
Shigeshi Omori
大森重志

Designer
Eiko Sakata
坂田瑛子

Bottle Designer
Toshihiko Daimon
大門俊彦

Client
Suntory Ltd.
サントリー株式会社
1972

244a

244b

In this mark of identity, the capital "S" in Suntory Foods has been made into an emblem resembling the G clef, the first and best-known symbol of musical notation. This dual significance stresses the harmony of quality, familiarity, and enjoyment in Suntory Foods. It also puts across a sense of the refreshment derived from these food products.
(Shigeshi Omori)

サントリーフーズの頭文字"S"をシンボライズするに当たって、清涼感はもとより、親しみやすさ、楽しさ、そして品質のハーモニーを強調すべく、音楽で親しみ深い五線譜の最初の記号"ト音記号"をモディファイしました。
(大森重志)

245
SUNTORY Foods Logo-Mark
サントリーフーズ ロゴマーク

Art Director & Designer
Shigeshi Omori
大森重志

Client
Suntory Ltd.
サントリー株式会社
1972

Suntory Foods

245

246
SUNTORY World Match Play Championship Logo-Mark
サントリーワールドマッチプレイ
選手権 ロゴマーク

Art Director
Shigeshi Omori
大森重志

Designer
Eiko Sakata
坂田瑛子

Client
Suntory Ltd.
サントリー株式会社
1974

suntory
WORLD MATCH PLAY
CHAMPIONSHIP

246

247a〜247h
Sinvino BI Plan
シンビーノBI計画

Art Director & Designer
Gan Hosoya
細谷 巖

Designer
Yasuko Kurakata
倉方康子

Client
Otsuka Foods Co., Ltd
大塚食品工業株式会社
1982

247a

In Italian, sin means "no," and vino means "wine." "Sinvino" is thus a new wine-type beverage that contains no alcohol. The simple label design uses the univers typeface and a tri-color scheme that divides the face of the label into blocks to create a youthful and Italianate image.
(Gan Hosoya)

イタリア語では「シン」はNO、「ビーノ」はワインのことである。アルコール分を含まない、ワインタイプの新しい飲料。ユニバースそのままのロゴタイプとトリコロール・カラーの単純な面分割で、若さとイタリア風のイメージを表現している。
(細谷 巖)

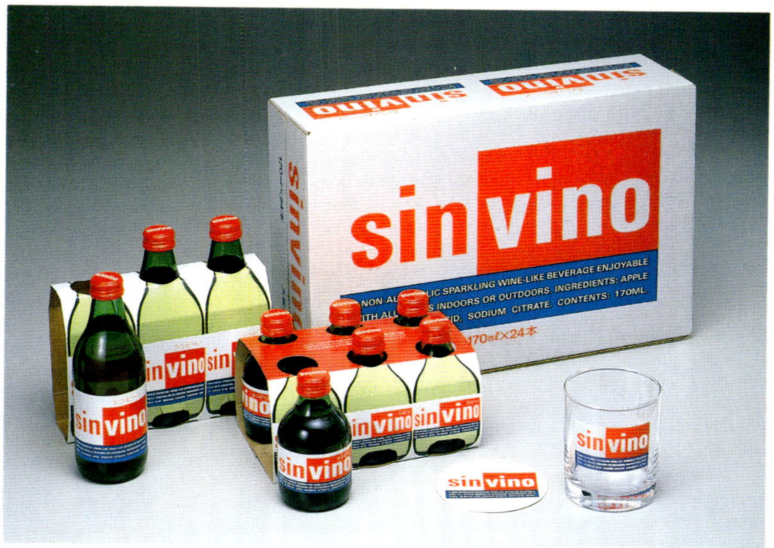

247b

247c

247d

247e

247f

247g

247h

248a〜248g
Calorie Mate BI Plan
カロリーメイトBI計画

Art Director & Designer
Gan Hosoya
細谷 巖

Logotype Designer
Tadasu Fukano
深野 匡

Designer
Yasuko Kurakata
倉方康子

Client
Otsuka Pharmaceutical Co., Ltd.
大塚製薬株式会社
1984

248a

This logo done in sporty, fluid lettering combines with the yellow ground and closely printed, full product description to project the image of a rich food product offering balanced nutrition. To maintain consistent visual appeal, the same design is carried through from the 200ml cans to cartons and novelties.
(Gan Hosoya)

スポーティで流動感のあるロゴタイプ、黄色一色の地色、ビッシリと組み込んだ説明文のスペース。すべて中味の濃いバランス栄養食品のイメージである。200ml缶からカートン、ノベルティに至るまで、徹底して同じデザインでまとめて総合的なアピールをねらった。
（細谷 巖）

248b

248c

248d

248e

248f

248g

249
OMNI Logotype
オムニ ロゴタイプ

Art Director
Gan Hosoya
細谷 巖

Designer
Tadasu Fukano
深野 匡

Client
Obunsha Co., Ltd.
株式会社旺文社
1982

オムニ 249

250
Niseko Golf Course Logo-Mark
ニセコゴルフコース ロゴマーク

Art Director & Designer
Gan Hosoya
細谷 巖

Client
Seibu Railway Co., Ltd.
西武鉄道株式会社
1986

Niseko
Golf Course 250

251
POCARI SWEAT Automat Symbol Mark
ポカリスエット 自動販売機 シンボルマーク

Art Director & Designer
Gan Hosoya
細谷　巖

Client
Otsuka Pharmaceutical Co., Ltd.
大塚製薬株式会社
1985

POCARI SWEAT 251

252
Magic Dragon Logo-Mark
マジックドラゴン ロゴマーク

Art Director
Gan Hosoya
細谷　巖

Symbol Mark Designer
Bou Numata
沼田　望

Logotype Designer
Tadasu Fukano
深野　匡

Client
Magic Dragon Japan Co., Ltd.
株式会社マジックドラゴン
1986

MAGIC DRAGON 252

253a〜253c
KIRICO BI Plan
キリコBI計画

Consulting
Hakuhodo Inc.
株式会社博報堂
Art Director
Masatoshi Toda
戸田正寿
Designer
Mamoru Suzuki
鈴木　守
Client
Suntory Ltd.
サントリー株式会社
1984

To point up the cool, crystalline image of vodka, the plane that forms the background for the lettering is divided into four segments. The colors blue, green, and red are used to symbolize the sky, the forest, and passion.
(Masayoshi Toda)

ウオッカの持つクールなイメージと透明感を生かすために、画面を4分割し、空と森と情熱の色を青、緑、赤、で構成しました。
(戸田正寿)

253a

253b

253c

254a〜254b
SUNTORY CAN BEER '83 BI Plan
サントリー カンビール '83 BI計画

Consulting
Hakuhodo Inc.
株式会社博報堂

Art Director
Masatoshi Toda
戸田正寿

Designer
Mamoru Suzuki
鈴木 守

Client
Suntory Ltd.
サントリー株式会社
1983

CAN SUNTORY 生BEER

254a

254b

We are living in an age when it is fashionable to buy canned beer by the carton. This trend inspired a stylish and distinctive carton design that captures the mood of modern apartment life.
(Masatoshi Toda)

「ダンボール箱ごと、缶ビールを買う時代」というマンションライフへの提案をコンセプトに、ダンボール箱のデザインをオシャレに個性的にしてみようと制作しました。
(戸田正寿)

255a〜255c
ISETAN CI Plan
伊勢丹CI計画

Art Director
Masatoshi Toda
戸田正寿

Designer
Mamoru Suzuki
鈴木 守

Client
Isetan Co., Ltd.
株式会社伊勢丹
1986

255a

The Isetan corporate identity scheme was developed using dark blue and yellow as the corporate colors. Dark blue has long been regarded as a basic color for clothing and other articles. Yellow, by contrast, is a typically modern color. The "I" in "Isetan" was initially made larger than the other letters solely for the sake of visual impact, but this feature later provided the basis for a clever advertising slogan. The logo was introduced to the public with the words "Ai ga ohkiku narimashita," which can mean both "Our 'I' has gotten larger" and "Our love has grown."
(Masatoshi Toda)

コーポレートカラーを、紺と黄にして展開。紺は昔からの着物などの基本の色。黄はモダンカラーの代表色と考えました。Iが大きくしてあるのは感覚的なものですが、後日、広告展開の中で「愛が大きくなりました。伊勢丹百貨店」とご紹介していただきました。
(戸田正寿)

255b

ISETAN

256
SUNTORY CAN BEER Logotype
サントリーカンビール ロゴタイプ

Consulting
Hakuhodo Inc.
株式会社博報堂

Art Director
Masatoshi Toda
戸田正寿

Designer
Mamoru Suzuki
鈴木 守

Client
Suntory Ltd.
サントリー株式会社

1983

256

257
SUIT Logotype
スーツ ロゴタイプ

Consulting
Dentsu Inc.
株式会社電通

Art Director
Masatoshi Toda
戸田正寿

Designer
Taku Nakamura
中村 卓

Client
D'urban Incorporated
株式会社ダーバン

1984

257

258
VIVRE 21 Logotype
ビブレ21 ロゴタイプ

Art Director
Masatoshi Toda
戸田正寿

Designer
Mamoru Suzuki
鈴木　守

Client
Harajuku Vivre 21
原宿ビブレ21
1984

VIVRE 21

258

259a〜259b
I. S. BI Plan
アイ・エスBI計画

Art Director & Designer
Kaoru Kasai
葛西　薫
Client
ISSEY MIYAKE INTERNATIONAL Inc.
株式会社イッセイミヤケインターナショナル
1983

I.S.
chisato tsumori design

259a

The "ISSEY SPORTS" line of fashions designed by Chisato Tsumori changed is name to simply "I.S." The combination of these initial letters already had the elements of a striking mark, but I added a color accent for the "I" to give it further effect. This mark was also used to adorn various types of apparel in the effort to speed up recognition of the I.S. brand.
(Kaoru Kasai)

津森千里氏デザインのファッションブランド、ISSEY SPORTSが、I.S.という名に変わった。IとSというアルファベットの組み合わせが、すでにマークとしての要素を備えており、Iに色のアクセントを与えたことで、より特長が得られた。衣服にも積極的にこのマークがデザインされ、I. S. ブランドの急速な浸透が図れた。
（葛西　薫）

259b

260a〜260b
CALL BI Plan
コールBI計画

Art Director
Kaoru Kasai
葛西　薫

Designer
Koichi Nishida
西田耕一

Client
SONY Corporation
ソニー株式会社
1985

CALL
Sony Shop

260a

With the advent of the audio-visual age, the electric shops found in local communities have had to double as AV shops. Call is the new unified name for Sony affiliated electric stores. The design communicates the sophistication, dignity and modernity typical of Sony. The exteriors of the shops, design of the sales areas, selection of goods, and approach to service were all changed and renewed in the pursuit of a richer, more fulfilled life —for regions, locales, and the shops themselves.
(Kaoru Kasai)

オーディオ・ビジュアル時代を迎え、いわゆる町の電気屋さんは、AV-SHOPへの返信が要求されてきた。CALLはそんなソニーショップのための統一店名である。ロゴタイプは、ソニーらしい知性、品位、モダンさなどをイメージし、また、外装、売場設計、商品選択、サービスなどすべて一新して、店、町、地域の生活のより豊かな変化を求めたものである。
（葛西　薫）

260b

261a〜261c
NUDE BI Plan
ヌードBI計画

Art Director & Designer
Kaoru Kasai
葛西　薫
Client
SONY Corporation
ソニー株式会社
1982

N·U·D·E
SONY STEREO EAR-RECEIVER

261a

Sony's new type of receiver, denuded of the headband that once connected headphones, is appropriately named "Nude." But the brand is actually an acronym standing for "Natural-and-Unequalled-Dynamic-Ear-receiver." I sought a presentation that would project the brand name to greatest effect. The logo is based in part on existing advertising designs, and simply a uses commercially available typeface.
(Kaoru Kasai)

従来のヘッドホーンからヘッドアームを取り去り、文字通り、ヌードとなったものが、このソニーレシーバー、N・U・D・Eであり、Natural and Unequalled Dynamic Earreceiverの品名でもある。
そんな商品名のインパクトが、できるだけ前に出るよう、また、従来の広告展開なども考慮して、ロゴタイプは凝らずに、既成書体の組み合わせにとどめた。
（葛西　薫）

261b

261c

262
Miscellanea Logotype
ミスレニア ロゴタイプ

Art Director
Kaoru Kasai
葛西 薫

Designer
Tetsuo Nishikawa
西川哲生

Client
The Seibu Department Stores, Ltd.
株式会社西武百貨店
1986

262

263
Bridal House BIBI Logotype
ブライダルハウスビビ ロゴタイプ

Art Director & Designer
Kaoru Kasai
葛西 薫

Client
Animal Park Co., Ltd.
アニマルパーク株式会社
1984

263

264
SOH Symbol Mark
双 シンボルマーク

Art Director & Designer
Kaoru Kasai
葛西 薫

Client
SOH Co., Ltd.
株式会社双
1984

264

265a〜265e
WAVE VI Plan
WAVE VI計画

Art Director
Kuni Kizawa
鬼澤 邦
Designer
Takashi Tanabe
田辺 卓
Client
WAVE ROPPONGI SEIBU
WAVE六本木西武
1983

WAVE
ROPPONGI·SEIBU

265a

"Wave" is a house of sound and vision whose striking achromatic image forms a chic contrast with the colorful background of Roppongi. The Wave logo—black lettering against a gray ground—complements the overall buidling design with angular lines and a metallic vibration.
(Kuni Kizawa)

音と映像の館"WAVE"。いろいろな情況から見て、有彩色を加味している六本木の地域に対して、無彩色(グレー、ブラック)の存在をイメージの基本とした。
基本色は無彩色のグレー、文字色は黒。ロゴタイプは鋭角的な線と金属的な響きのあるイメージで発想。
(鬼澤 邦)

265b

265c

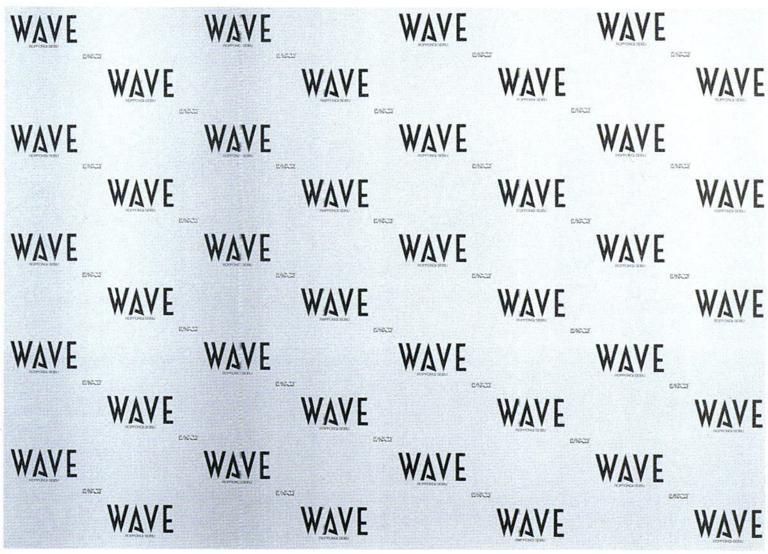

265d

265e

266a〜266b
JUN MEN VI Plan
ジュン・メンVI計画

Art Director & Designer
Makoto Saito
サイトウ・マコト

Designer
Minoru Kanda
神田みのる

Photographer
Tsutomu Wakatsuki
若月 勤

Client
Jun Co., Ltd.
株式会社ジュン
1982

266a

266b

This logo design was commissioned for the company's renewal campaign in 1982. Jun's former logo was delicate and refined, rather reminiscent of medieval Europe. For the new image we chose a typographic design that is contemporary and masculine.
(Makoto Saito)

'82にリニュアルキャンペーンに向けてのゴロタイプデザインの変更を依頼された。従来のJUNのロゴタイプは繊細でヨーロッパの中世を匂わすものであった。これに対して現代的で男らしさの匂うタイポデザインにした。
(サイトウマコト)

267a~267b
GARO VI Plan
ガロ洋装店VI計画

Art Director
Makoto Saito
サイトウ・マコト

Designer
Soko Hosonome
細目壮子

Photographer
Hiroshi Yoda
与田弘志

Client
Jun Co., Ltd.
株式会社ジュン
1984

GARO
洋装店

267a

267b

Garo is another Jun Co., Ltd. brand. The former logos for those Garo and Jun were of a European medieval style. For the new image we sought a somewhat playful design aimed at the young generation.
(Makoto Saito)

㈱JUNの一つのブランドである。このガロもJUNと同じくヨーロッパの中世のテクスチャーを生かしたデザインであったが、これを嫌って若者層を狙った若干遊び心を加えたデザインにした。
(サイトウマコト)

268a〜268b
TO VI Plan
ティーオーVI計画

Art Director
Makoto Saito
サイトウ・マコト

Designer
Soko Hosonome
細目壮子

Client
Jun Co., Ltd.
株式会社ジュン
1985

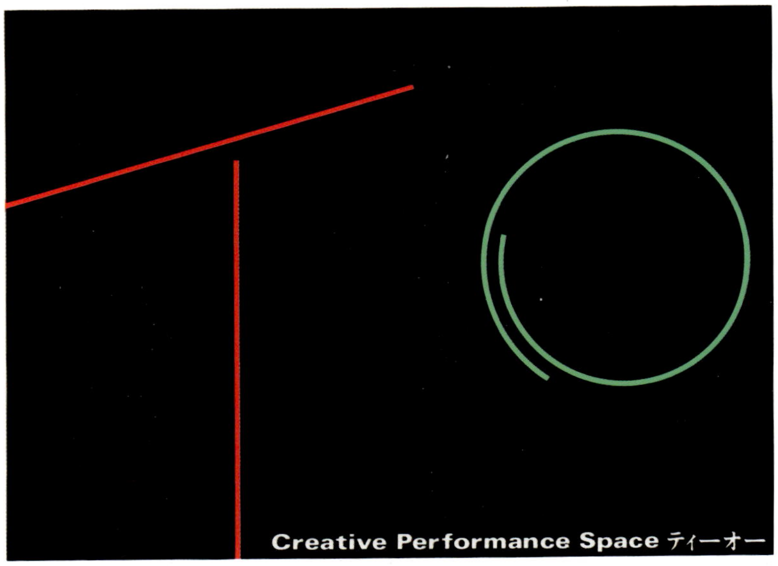

268a

268b

To [ti:óu] is a building complex for fashion shops of the Jun chain. Its name is simply the English "to" intended in the sense of "toward something, somewhere." The black background and fine white lines of the letters were designed to create a logo suited to this image.
(Makoto Saito)

TO は㈱JUN の一環したファッションテナントビルの集合体である。TO［ティーオー］というのは英語の "to" の「どこどこへ」という不定形な意味である。これにあわせてロゴは真黒な空間に白い筋が文字を描いているかの様なデザインにした。
(サイトウマコト)

269a〜269b
IDEX VI Plan
イデックスVI計画

Creative Director
Michio Akiyama
秋山道男

Art Director
Makoto Saito
サイトウ・マコト

Designer
Soko Hosonome
細目壮子

Photographer
Tsutomu Wakatsuki
若月　勤

Client
Idex Co., Ltd.
株式会社イデックス
1984

269a

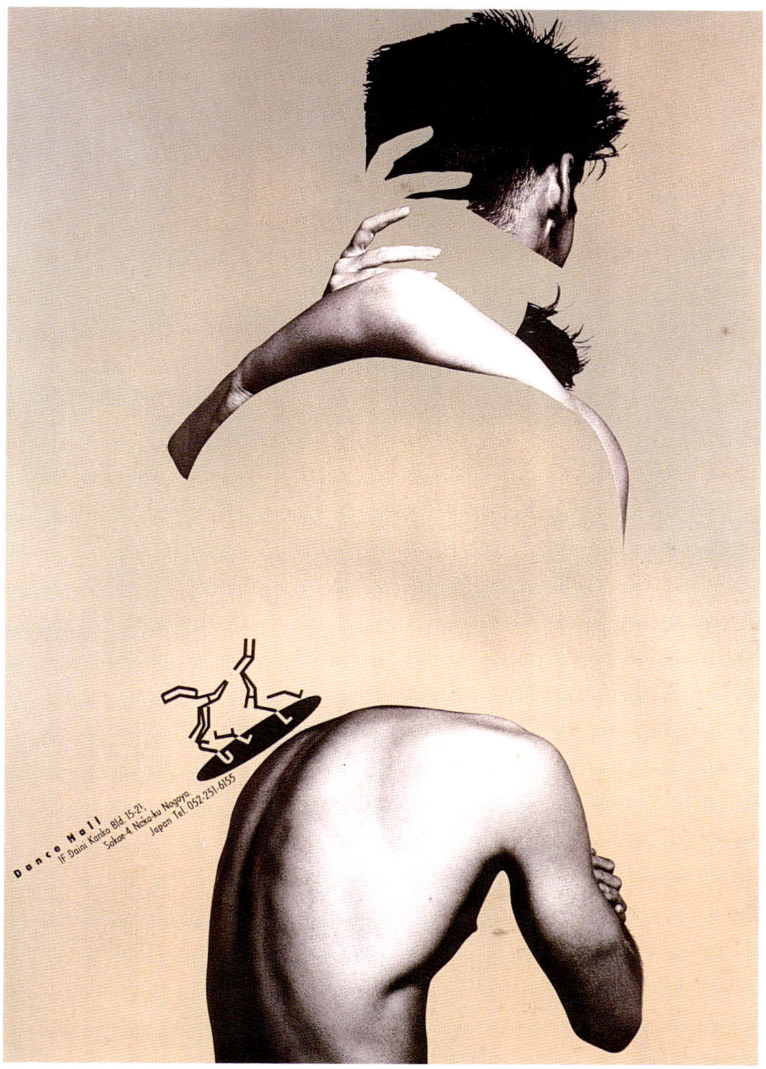

269b

This typographic design is for a disco, as the words themselves suggest, and the letters are made to come alive, literally dancing in the spotlight.
(Makoto Saito)

これは読んで字のごとくディスコホールのタイポデザインである。文字を人間がダンシングしているかの様な少し動きのある形にし、それにスポットが当たっているかの様なデザインにした。
(サイトウマコト)

270a〜270g
Hôtel de Mikuni VI Plan
オテル・ドゥ・ミクニVI計画

Art Director & Designer
Douglas Doolittle
ダグラス・ドリトル
Interior Designer
Uni Plan Inc.
株式会社ユニ・プラン
Client
Hôtel de Mikuni
オテル・ドゥ・ミクニ
1986

270a

It was Christmas Day when I made the first presentation to Mr. Mikuni, a man of artistic detail but International standing. His widely acclaimed French restaurant needed a creative identity that reflected such stature. Based on his signature I devised a logo that was asthetic, ageless and ever flowing throughout all its applications be they menus, wall reliefs, tableware, or signage.
Thank you Mr. Mikuni. A client I have never had to work for but alwats with…
(Douglas Doolittle)

私が初めて三国氏に会ってプレゼンテーションをしたのは、クリスマスの日だった。三国氏は審美的な繊細さを持ち、国際的な視座に立っている人である。彼は有名なフランス料理のレストランの格付けにふさわしいアイデンティティを求めていた。私は彼の署名をもとにして、美しく、すたれず、しかもメニューや壁のレリーフ、食器などに幅広く使えるものを考えた。三国氏に感謝したい。ともに仕事をするクライアントとしての彼に…。
(ダグラス・ドリトル)

270b

270c

270d

270e

270f

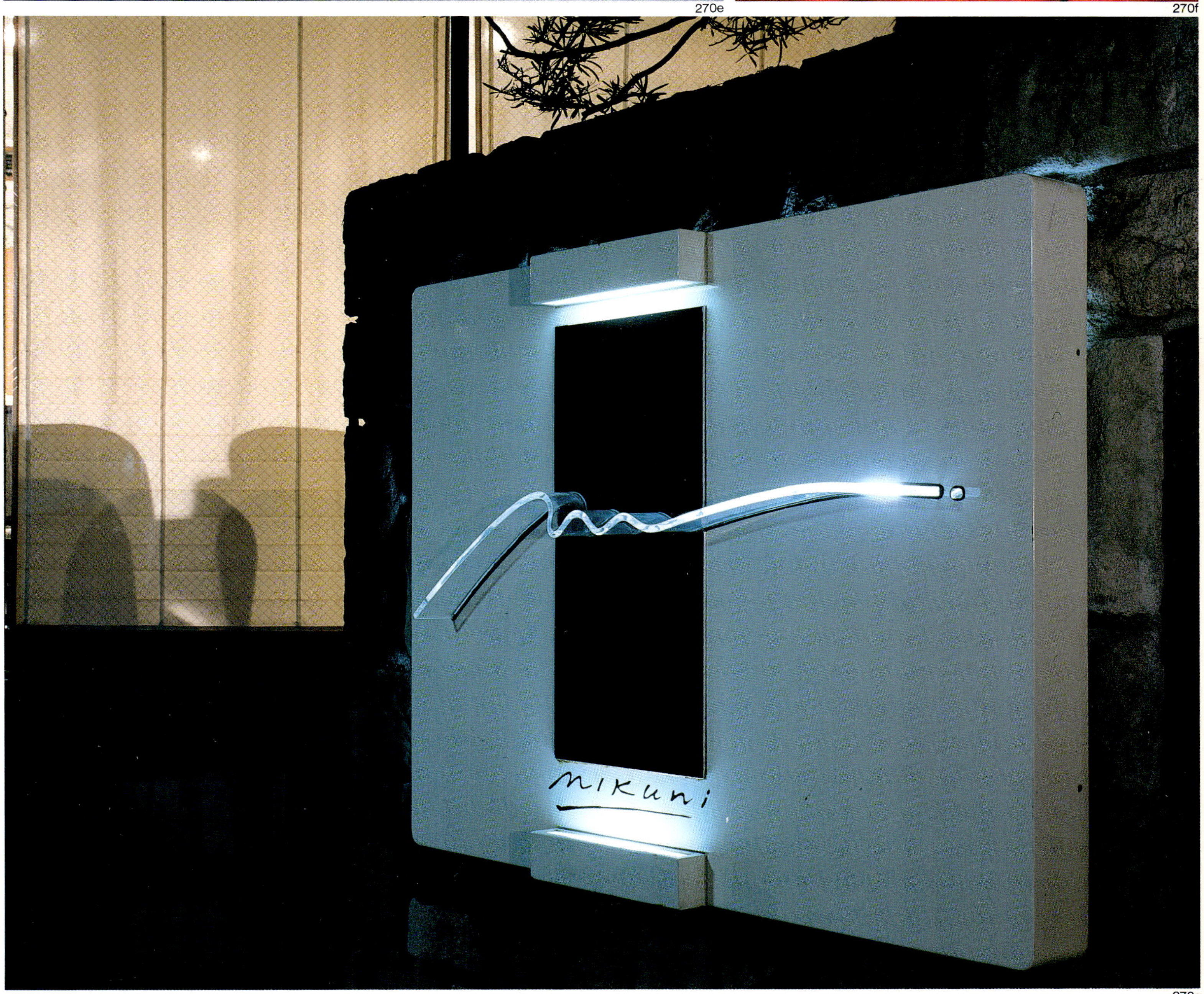

270g

271a

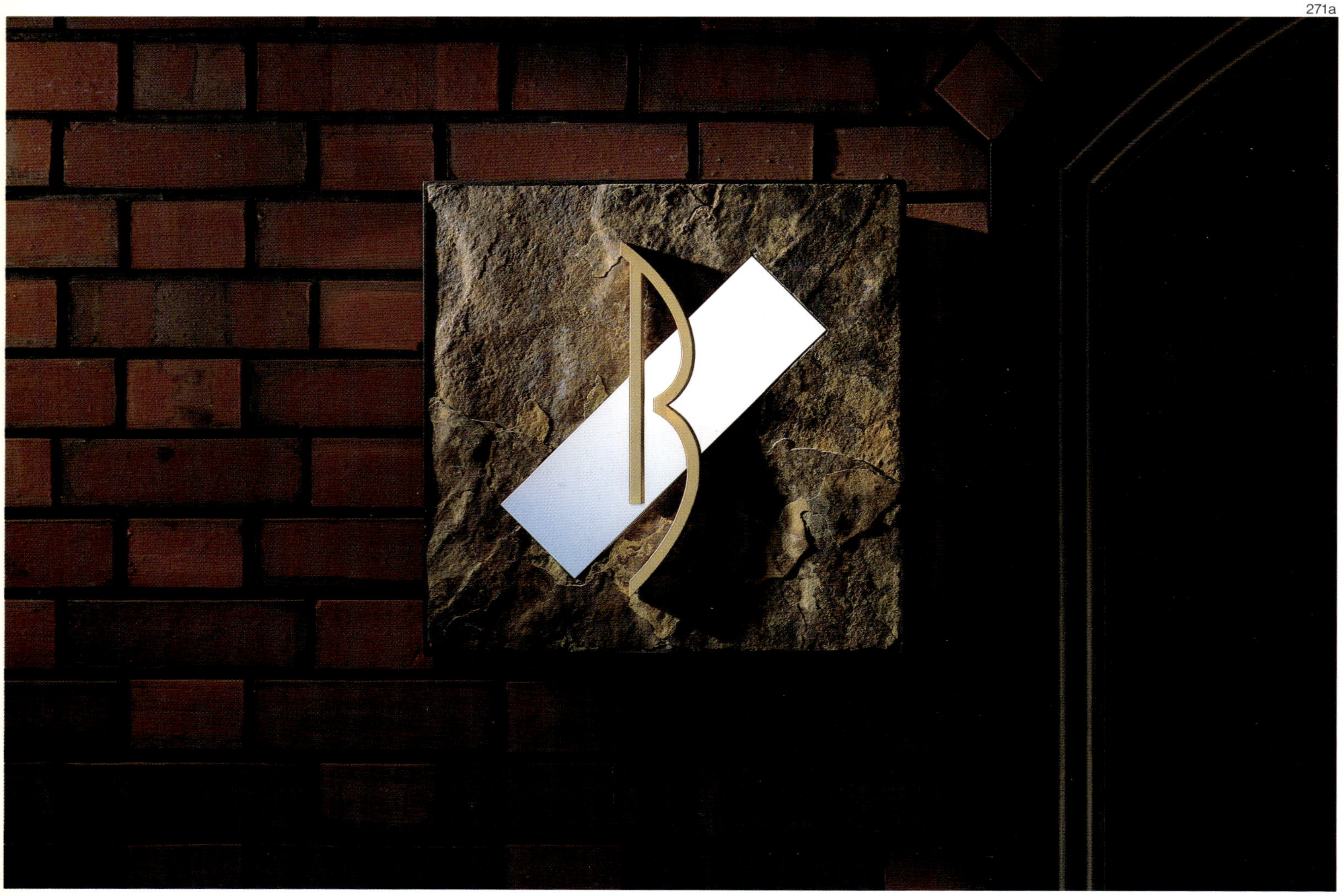

271b

271a〜271d
Bloomin' Bar VI Plan
ブルーミンバーVI計画

Art Director & Designer
Douglas Doolittle
ダグラス・ドリトル
Client
Bloomin' Bar
ブルーミン バー
1987

271c

The concept behind this project was to devise a identity for a bar that would appeal to an exclusive, creative, adult audience in central Harajuku.
It was not a members club but needed that "exclusive" image to discourage flightly customers. To subconsciously assist in this screening process the name, Bloomin' Bar was used only in selected visual applications so new clientelle had to be personally introduced to the establishment.
The simplicity of the stylized "B" in contrast to the rust-accentuated Italian state slab forms a sophisticated, yet dynamic image.
(Douglas Doolittle)

このプロジェクトのコンセプトは、原宿界隈に出かけてくるエグゼクティヴでクリエイティヴな大人のためのバーの、アイデンティティをつくり出し、会員制のクラブではないが、一見の客が入りにくいようにエグゼクティヴ風にする必要があった。
ブルーミン・バーという名前は、ごく限られたところにしか用いられていないので、新しい客は個人的に紹介されないとこの場所に来られない。しかしそれは、気づかれないようにして客層を選別するためなのである。
赤サビ色のイタリアンスレートスラブとは対照的に、デザイン化された "B" の文字の形の単純性が、洗練され、いっそうダイナミックなイメージを構成している。
(ダグラス・ドリトル)

271d

272a〜272d
SEED CI Plan
シード CI計画

Consulting
Seiichi Mizuno
水野誠一

Art Director & Designer
Tsuguya Inoue
井上嗣也

Designer
Beans
株式会社ビーンズ

Client
The Seibu Depertment Stores, Ltd.
株式会社西武百貨店
1986

SEED
SHIBUYA SEIBU
272a

This is a splendid name. The meaning and information carried by the word "seed" is multifarious and strong. An organic, primitive touch was desirable for the lettering, I thought. The photograph for the posters is of seeds, a literal portrayal of the store's name and a current hobby of mine. The seeds photographed are selections from a collection of seeds from North America, Mexico, and Panama. (Tsuguya Inoue)

ネーミングが素敵だ。SEEDという言葉の持っている意味と情報は既に多く、強い。ロゴタイプに関しては有機的でプリミティブなものでありたいと思った。ポスターの映像はネーミングそのまま、私の興味の真只中の「種」である。アメリカ、メキシコ、パハマ他のそれらの中から選んでつくりました。
(井上嗣也)

272b

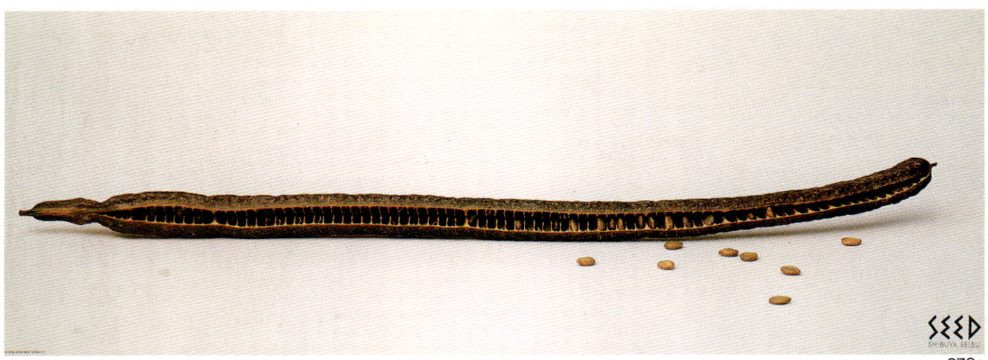

272c

272d

273a〜273j
The Weekly Fluctuant Book Logotype
週刊本 ロゴタイプ

Creative Director
Mikitaka Nakano
中野幹留

Art Dirctor & Designer
Tsuguya Inoue
井上嗣也

Designer
Beans
株式会社ビーンズ

Client
Asahi Press
株式会社朝日出版社
1984

273b

273c

週刊本
THE WEEKLY FLUCTUANT BOOK
273a

273d

273e

273f

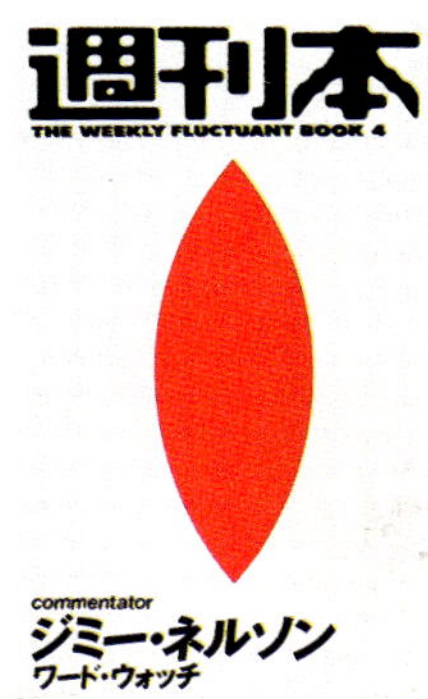

273g

273h

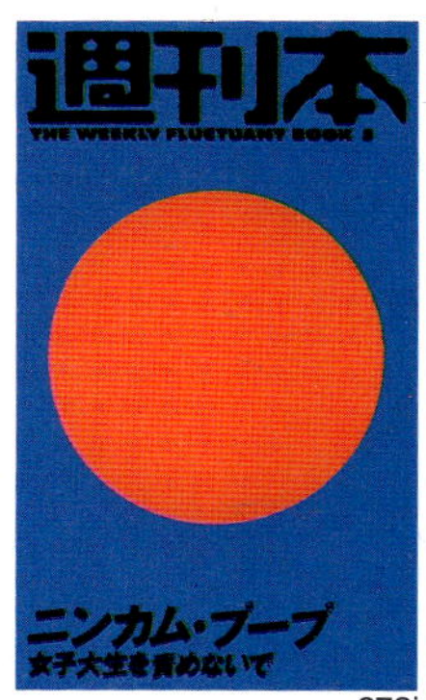

273i

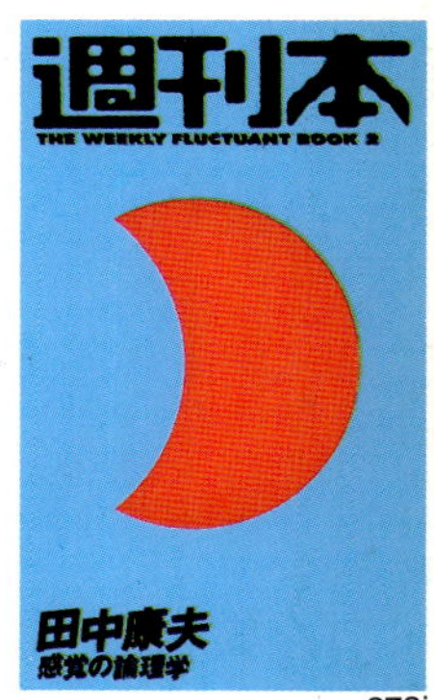

273j

274a〜274i
JR Group CI Plan
JRグループCI計画

Consulting
Dentsu Inc.
株式会社 電通
Nippon Design Center
株式会社日本デザインセンター

Art Director & Designer
Yoji Yamamoto
山本洋司

Client
JR Group
JRグループ
1987

JR

274a

JR北海道

JR東日本

JR東海

JR西日本

JR四国

JR九州

JR貨物

JRグループ

274b

Since JR is a public transportation system, I aimed to create a logo design which reflected this public character and would continue on in tireless service for years like the corporation itself. I sought a design that was international in style, refined in form, and easily understood and recognized by people of all ages. Since it would be emblazoned on trains, it could not be oriented to either right or left. (Yoji Yamamoto)

輸送機関であるので公共性が強いこと、何年経過しても飽きられず長続きすること、国際性があること、子供からお年寄りまで誰が見ても理解出来親しまれること、造形的な完成度が高いこと、列車に付けるので一方だけの方向性があってはならないJRの文字が右、左、に動いても違和感のないこと、など考えてデザインしています。
(山本洋司)

274c

274d

274e

274f

274g

274h

274i

275a〜275c
Iichiko Logotype
いいちこ ロゴタイプ

Art Director
Hideya Kawakita
河北秀也

Logotype Designer
Hidetaka Myojyo
明星秀隆

Client
Sanwa Shurui Co., Ltd.
三和酒類株式会社
1983

iichiko

275a

275b

275c

This logo for a brand of shochu, a favorite distilled liquor, seeks a Japonesque flavor using the English romanization. The layout is very simple to facilitate all kinds of variations: open or tight letter space, italic type or vertical setting. I thought using the natural print characters would actually have a greater impact than trying to create a distinctive typeface. (Hideya Kawakita)

「いいちこ」の英文字のロゴタイプは、日本の古来からの焼酎ということもあり、できるだけ、英文でありながら日本的に感じられるようにデザインした。さらに、字間を広げたり、狭くしたり、斜体をかけたり、タテ組にした時も自由に使えるように、単純な形にした。長く使われるロゴタイプは、あまり特徴を持ったものより、自然な書体の方がよい印象を与えるのではないかと思う。
(河北秀也)

276a〜276b
EX VI Plan
エックスVI計画

Art Director
Keisuke Nagatomo
長友啓典

Designer
Ichiro Masuzumi
増住一郎

Interior Designer
Takashi Sugimoto
杉本貴志

Copy Writer
Shinzo Higurashi
日暮真三

Client
Toku-Ju Co., Ltd.
株式会社徳寿
1984

276a

276b

Both the sophisticated interior design by Sugimoto Takashi and the name and its sound of the word by Shinzo Higurashi, along with the location of the shop on Shibuya's Koen Dori, helped greatly to develop this design. Since I was also in charge of the graphics used for the Junikagetu Building in which the brasserie "EX" is housed, this project was a comparatively easy one to complete.
(Keisuke Nagatomo)

杉本貴志氏によるハイタッチなインテリアデザイン、日暮真三氏によるネーミングとその響き、そして渋谷の公園通りという立地によって、このロゴのデザインはほぼ決定した。また、このブラッセリー"EX"が入っている"12ヶ月"というビル全体のグラフィックデザインを受け持ったことによって、比較的仕事がやりやすかった。
(長友啓典)

277a〜277c
IMA-Hikarigaoka Life Station VI Plan
IMA-光が丘ライフステーション VI計画

Art Director
Yoshiaki Okura
大蔵良章

Art Director & Designer
Katsuichi Ito
伊藤勝一

Designer
Katsumi Miyasaka
宮坂克己

Client
Hikarigaoka Toshi Center Co., Ltd.
株式会社光が丘都市センター
1987

The logo for the IMA Hikarigaoka Life Station has as its motif the IMA of the center's name. IMA Hikarigaoka is not just a shopping center, and its compound character as a "life station" combining shopping with sports and cultural facilities as suggested by the design created as if built with a child's building blocks. The letter "M" is made in the shape of a box with an open top to impart the pleasant sense of expectation that something exciting may pop out at any minute.
(Katsuichi Ito)

ネーミングのIMAをモチーフとしたシンボルマーク。単なるショッピングセンターではなく、ライフステーションとしてスポーツ・文化施設を併せ持つ複合的性格を「積み木の組み合わせ」でシンボライズしたものである。特に「M」の字は箱の口が開いた形で、楽しい刺激が飛び出してくる期待感を表現している。
(伊藤勝一)

277a

277b

277c

278

279

280

281

278
Asahi Chemical Industry Co., Housing Administration Symbol Mark
旭化成工業住宅事業部門 シンボルマーク

Consulting
Kurisen Ltd.
株式会社クリセン

Art Director
Katsuichi Ito
伊藤勝一

Designer
Katsumi Miyasaka
宮坂克己

Client
Asahi Chemical Industry Co., Ltd.
旭化成工業株式会社
1982

279
Kokusai Securities Symbol Mark
国際証券 シンボルマーク

Art Director
Yukioki Hasegawa
長谷川幸興

Art Director & Designer
Katsuichi Ito
伊藤勝一

Client
Kokusai Securities Co., Ltd.
国際証券株式会社
1983

280
Nikko Hotels International Symbol Mark
ニッコー ホテルズ インターナショナル シンボルマーク

Art Director & Designer
Katsuichi Ito
伊藤勝一

Client
Japan Air Lines Development Co., Ltd.
日本航空開発株式会社
1980

281
DIY Logo-Mark
DIY ロゴマーク

Art Director
Katsuhiko Yasuda
安田勝彦

Designer
Katsuichi Ito
伊藤勝一

Client
Hinode Corporation
ヒノデ株式会社
1981

282a〜282d
Tobu Department Stores Logo-Mark
東武百貨店ロゴマーク

Art Director & Designer
Kakutaro Iimori
飯守格太郎
Client
Tobu Department stores Co., Ltd.
株式会社東武百貨店
1983

282a

282b

282c

A recent trend in pronunciation among young Japanese is to drop the final vowel of words, a habit that produces a clipped, rapid pattern of speech. The linking verb desu, for example, usually enunciated in two syllables, is now pronounced "des." We omitted the "u" of Tobu to reflect this trend in contemporary speech. The other feature of the design is elongation of the tail of the "o," a treatment that gives graphic expression to the long sound of the vowel, and evokes a visual rhythm in the symbol.
(Kakutaro Iimori)

TOBUのUを除去した。近頃の日本人の発音は、母音を省略して子音だけですますことにより、めりはりのはっきりした、速度の速い発音、たとえば「…です」は〈desu〉でなく、〈des〉となりつつある。もうひとつは、〈●●●●〉から〈●●-●〉へと音節をひとつにしながらも、ひとつのリズムのある記号性を視覚的に持たせた。
(飯守格太郎)

282d

283a〜283b
Shisaku Nonsense literature Symbol Mark
思索ナンセンス選集 シンボルマーク

Art Director & Designer
U. G. Sato
U. G. サトー
Client
Shisakusha Publishing Co., Ltd.
株式会社思索社
1982

283a

283b

This is the symbol created for a selected collection of nonsense literature by eight authors planned by the Shisakusha publishing company. It depicts a reader and a nonsensical "flight of the imagination."
(U.G. Sato)

思索社が企画した８人の作家による「ナンセンス選集」のシンボルマーク。本と、読む人に加えて、ナンセンスな飛躍のイメージを盛りこみながら、デザインした。
(U. G. サトー)

284a〜284c
Pyramid Zone VI Plan
ピラミッドゾーンVI計画

Art Director & Designer
Takayuki Soeda
副田高行
Client
Iwataya Department Store Co., Ltd.
株式会社岩田屋
1984

284a

The mark is an unassuming, boldly executed but simple pyramid. In order to merge the typography with the mark, I created an inorganic lettering. Of course, it is relatively easy to make a striking mark for a corporation whose name evokes a specific graphic image. In the case of a pyramid, indeed, the shape itself is a forceful one that possesses a mystical quality all its own.
(Takayuki Soeda)

マークは、素直に、強くてシンプルなピラミッドを視覚化しました。ロゴは、マークと一体感のある無機的な文字をつくりました。ピラミッドゾーンというような、具体的なイメージをもっているネーミングのマークづくりは、比較的やりやすいですね。特に、ピラミッドのように、そのものじたいが非常にすぐれた造形力と、神秘性をもっているものはそうですね。
(副田高行)

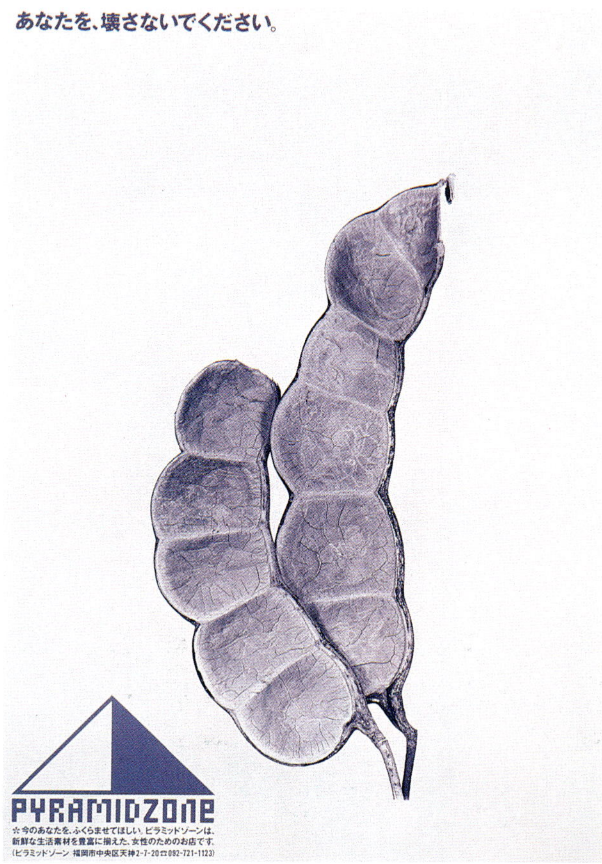

284b

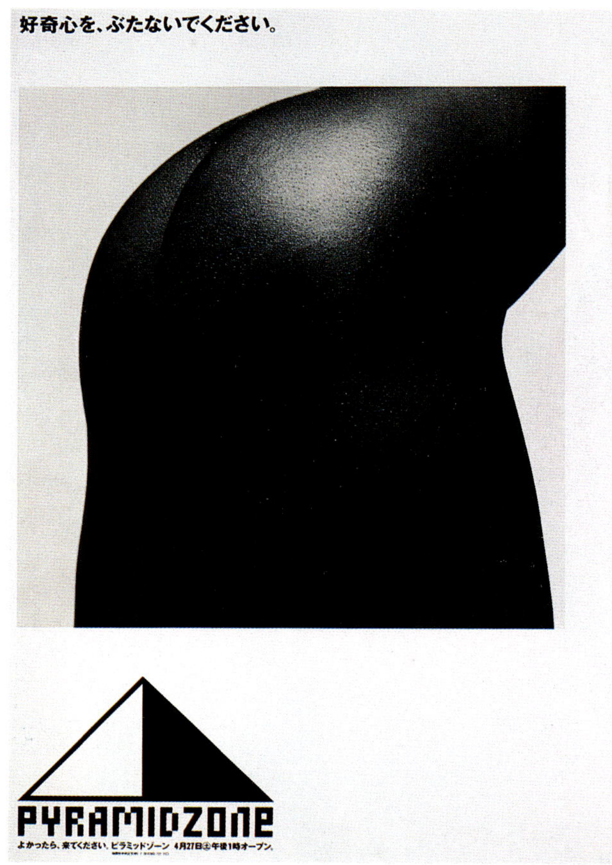

284c

285
Yeellow Magic Orchestra Symbol Mark
イエローマジックオーケストラ シンボルマーク
Art Director & Designer
Yukimasa Okumura
奥村靫正
Client
Yoroshita Music, Inc.
株式会社ヨロシタミュージック
1981

YELLOW MAGIC ORCHESTRA 285

286
SCHOOL Label Logo-Mark
スクールレーベル ロゴマーク
Art Director & Designer
Yukimasa Okumura
奥村靫正
Client
MDI Inc.
株式会社ミディ
1984

SCHOOL 286

287
Littai Production Logo-Mark
立体プロダクション ロゴマーク
Art Director & Designer
Yukimasa Okumura
奥村靫正
Client
Littai Production Co., Ltd.
立体プロダクション株式会社
1983

LITTAI 287
PRODUCTION

288
MEDIA VIEW Symbol Mark
メディア・ビュー シンボルマーク

Art Director
Katsumi Asaba
浅葉克己

Designer
Valentina Grego
ヴァレンティーナ・グレゴ

Client
Media View Inc.
株式会社メディア・ビュー
1985

MEDIA VIEW INC.

288

289
Tokyo Itoi Shigesato Office Symbol Mark
東京糸井重里事務所 シンボルマーク

Designer
Katsumi Asaba
浅葉克己

Client
Tokyo Itoi Shigesato Office
東京糸井重里事務所
1979

ITOI
shigesato

289

290
Seibu Sports Symbol Mark
西武スポーツ シンボルマーク

Art Director
Katsumi Asaba
浅葉克己

Designer
Mitsunobu Uesaka
上坂光信

Client
The Seibu Depertment Stores, Ltd.
株式会社西武百貨店
1981

SEIBU SPORTS

290

291
CMC Logotype
シーエムシー ロゴタイプ

Art Director
Katsumi Asaba
浅葉克己

Designer
Noboru Hori
堀 昇

Client
Copywriters' Membership Club.
コピーライターズ メンバーシップ クラブ
1977

291

292
MOTHERS Logo-Mark
マザース ロゴマーク

Art Director
Katsumi Asaba
浅葉克己

Designer
Masato Oki
大木理人

Client
Mothers Inc.
株式会社マザース
1982

292

293
Yamagiwa Need Logo-type
ヤマギワ Need ロゴタイプ

Art Director
Katsumi Asaba
浅葉克己

Designer
Homare Hino
日野 誉

Client
Yamagiwa Interior Development Co., Ltd.
株式会社ヤマギワインテリヤ開発
1976

293

294
Hanae Mori Tights Symbol Mark
ハナエ・モリ タイツ シンボルマーク
Art Director
Katsumi Asaba
浅葉克己
Designer
Harumi Yamaguchi
山口はるみ
Client
Hanae Mori Tokyo Co.
株式会社ハナエ・モリ
1979

294

295a〜295b
Hanae Mori Private Symbol Mark
ハナエ・モリ プライベート シンボルマーク
Art Director
Katsumi Asaba
浅葉克己
Designer
Masaharu Takada
高田正治
Client
Nishikawa Sangyo Co., Ltd.
西川産業
1985

295a

295b

296
Tokyo Designers Space Logo-Mark
東京デザイナーズ・スペース ロゴマーク
Designer
Katsumi Asaba
浅葉克己
Client
Tokyo Designers Space
東京デザイナーズ・スペース
1977

296

297
Aoi Advertising Promotion Logotype
葵プロモーション ロゴタイプ

Art Director
Katsumi Asaba
浅葉克己

Designer
Masami Yamamoto
山本昌美

Client
Aoi Advertising Promotion Inc.
株式会社葵プロモーション
1986

297

298
TV eye Symbol Mark
ティーヴィー・アイ シンボルマーク

Art Director
Katsumi Asaba
浅葉克己

Designer
Masato Oki
大木理人

Client
Nippon Television Network Corporation
日本テレビ放送網株式会社
1980

298

299
KING KONG Logo-Mark
キングコング ロゴマーク

Designer
Katsumi Asaba
浅葉克己

Client
Ping Pong Team of Tokyo King Kong
卓球チーム キングコング
1976

299

300
KANSAI TELECASTING
Symbol Mark
関西テレビ シンボルマーク

Art Director & Designer
Yoshio Hayakawa
早川良雄

Client
Kansai Telecasting Corporation
関西テレビ放送株式会社
1963

300

301
Mishoryu Nakayama
Bumpo Kai Symbol Mark
未生流中山文甫会 シイボルマーク

Art Director & Designer
Yoshio Hayakawa
早川良雄

Client
Mishoryu Nakayama Bumpo Kai
未生流中山文甫会
1957

301

302
Wacoal Interior Fabric
Logo-Mark
ワコール・インテリアファブリック ロゴマーク

Art Director & Designer
Yoshio Hayakawa
早川良雄

Client
Wacoal Corp.
株式会社ワコール
1973

302

303

304

305

306

307

308

303
Tenpura-Shori Logotype
天ぷら松利 ロゴタイプ
Art Director & Designer
Yoshio Hayakawa
早川良雄
Client
Tenpura-Shori
天ぷら松利
1960

304
NANCY Logo-Mark
ナンシー ロゴマーク
Art Director & Designer
Yoshio Hayakawa
早川良雄
Client
Glass Gallery Nancy
グラスギャラリー・ナンシー
1970

305
Shiki Fabric House Symbol Mark
四季ファブリックハウス シンボルマーク
Art Director & Designer
Yoshio Hayakawa
早川良雄
Client
Shiki Fabric House Co., Ltd.
株式会社四季ファブリックハウス
1962

306
Gallery-Imai Symbol Mark
ギャラリー今井 シンボルマーク
Art Director & Designer
Yoshio Hayakawa
早川良雄
Client
Gallery-Imai
ギャラリー今井
1972

307
Club-Ido Symbol Mark
クラブ・イド シンボルマーク
Art Director & Designer
Yoshio Hayakawa
早川良雄
Client
Club-Ido
クラブ・イド
1960

308
Kanagawa Design Society Symbol Mark
神奈川県デザイン協会 シンボルマーク
Art Director & Designer
Yoshio Hayakawa
早川良雄
Client
Kanagawa Design Society
神奈川県デザイン協会
1976

309
Canon Logotype
キャノン ロゴタイプ

Designer
Kenji Ito
伊藤憲治

Client
Canon Inc.
キャノン株式会社
1954

309

310
Aichi Logotype
アイチ ロゴタイプ

Designer
Kenji Ito
伊藤憲治

Client
Aichi Co., Ltd.
株式会社アイチ
1973

310

311
Mimuro Logo-Mark
ミムロ ロゴマーク

Art Director
Eita Shinohara
篠原栄太

Designer
Kenzo Nakagawa
中川憲造
Hiroyasu Nobuyama
延山博保

Client
Mimuro Co., Ltd.
株式会社ミムロ
1986

312
PANDA Logotype
パンダ ロゴタイプ

Art Director
Kenzo Nakagawa
中川憲造

Designer
Hiroyasu Nobuyama
延山博保

Client
WWF "World Wide Fund for Nature Japan"
財団法人世界野生生物基金日本委員会
1985

313
ONKYO Symbol Mark
オンキョー シンボルマーク

Art Director
Hiroyuki Machida
町田浩之
Kenzo Nakagawa
中川憲造

Designer
Hiroyasu Nobuyama
延山博保

Client
Onkyo Corporation
オンキョー株式会社
1986

311

312

313

314
Noir et Blanc Symbol Mark
ノワールェブラン シンボルマーク

Art Director
Hirosi Kojitani
麹谷 宏

Designer
Kensuke Irie
入江健介

Client
Noir et Blanc
ノワールェブラン
1983

314

315
California Wine Symbol Mark
カリフォルニアワイン シンボルマーク

Art Director
Hirosi Kojitani
麹谷 宏

Designer
Kensuke Irie
入江健介

Client
California Wine Association
カリフォルニアワイン協会
1986

315

316
Honda Architect Symbol Mark
本田設計 シンボルマーク

Art Director
Hirosi Kojitani
麹谷 宏

Designer
Kensuke Irie
入江健介

Client
E, Honda Architect and Engineers'
本田設計事務所
1978

316

317

318

319

320

317
Mikami Symbol Mark
ミカミ シンボルマーク
Art Director
Hirosi Kojitani
麹谷 宏
Designer
Kensuke Irie
入江健介
Client
Mikami Co., Ltd.
株式会社ミカミ
1981

318
Bistro Sanno Symbol Mark
ビストロ・サンノーシンボルマーク
Art Director
Hirosi Kojitani
麹谷 宏
Designer
Kensuke Irie
入江健介
Client
Bistro Sanno
レストランビストロ・サンノー
1986

319
A Wood of Bach Symbol Mark
バッハの森 シンボルマーク
Art Director
Hirosi Kojitani
麹谷 宏
Designer
Kensuke Irie
入江健介
Client
Cultural Foundation for Bach Memorial Forest in Tsukuba
財団法人筑波バッハの森文化財団
1984

320
Japan Confectionery Fair Symbol Mark
菓子博覧会 シンボルマーク
Art Director
Hirosi Kojitani
麹谷 宏
Designer
Kensuke Irie
入江健介
Client
Japan Confectionery Fair
全国菓子大博覧会
1983

321
Tewaza Logo-Mark
テワザ ロゴマーク

Designer
Koichi Sato
佐藤晃一

Client
Fukutake Publishing Co., Ltd.
株式会社福武書店
1985

321

322
KIREN Logo-Mark
気練 ロゴマーク

Designer
Koichi Sato
佐藤晃一

Client
Office KIREN
オフィス気練
1985

322

323
Seiseian Logo-Mark
青々庵 ロゴマーク

Designer
Koichi Sato
佐藤晃一

Client
Shinso Ohasi Co., Ltd.
株式会社新装大橋
1986

323

324
Yasutomo Studio Symbol Mark
安友写真事務所 シンボルマーク
Art Director & Designer
Tetsuya Ota
太田徹也
Client
Yasutomo Studio
安友写真事務所
1985

324

325
CAFÉ BAR GREEN Logotype
カフェバー グリーン ロゴタイプ
Art Director & Designer
Tetsuya Ota
太田徹也
Client
Café Bar Green
カフェバー グリーン
1986

325

326
K—Publicity Symbol Mark
ケー・パブリシティ シンボルマーク
Art Director & Designer
Tetsuya Ota
太田徹也
Client
K—Publicity
ケー・パブリシティ
1979

326

327
Achilles Logotype
アキレス ロゴタイプ

Designer
Tadasu Fukano
深野 匡

Client
Achilles Corporation
アキレス株式会社
1977

327

328
NASCO Logo-Mark
日産航空サービス ロゴマーク

Art Director & Designer
Tadasu Fukano
深野 匡

Symbol Mark Designer
Nobol Okata
大方 昇

Client
Nissan Air Service Co., Ltd.
日産航空サービス株式会社
1983

328

329
JRCS 100 Anniversar Symbol Mark
日赤創立100周年 シンボルマーク

Designer
Tadasu Fukano
深野 匡

Client
Japanese Red Cross Society
日本日赤十字社
1978

329

330
Tokyo Telecommunication Network Symbol Mark
東京通信ネットワーク シンボルマーク

Art Director & Designer
Masuteru Aoba
青葉益輝

Client
Tokyo Telecommunication Network Company Inc.
東京通信ネットワーク
1986

330

331
International Education Center Symbol Mark
東京教育振興会 シンボルマーク

Art Director & Designer
Masuteru Aoba
青葉益輝

Client
International Education Center
財団法人国際教育振興会
1985

331

332
3rd International Design Festival Symbol Mark
第3回国際デザインフェスティバル
シンボルマーク

Art Director & Designer
Masuteru Aoba
青葉益輝

Client
Japan Design Foundation
財団法人国際デザイン交流協会
1986

332

333
BENGARAYA Symbol Mark
べんがらや シンボルマーク

Designer
Akisato Ueda
上田晃郷

Client
Sendan Co., Ltd.
株式会社せんだん
1979

334
HARMAN Symbol Mark
ハーマン シンボルマーク

Art Director
Kyosuke Yoshino
芳野恭輔

Designer
Akisato Ueda
上田晃郷

Client
Harman Co., Ltd.
株式会社ハーマン
1986

335
Drive-In Azuma Symbol Mark
ドライブイン アズマ シンボルマーク

Art Director & Designer
Akisato Ueda
上田晃郷

Client
Drive-In Azuma
ドライブ イン アズマ
1975

336
Idea Logotype
Idea ロゴタイプ

Designer
Akisato Ueda
上田晃郷

Client
PHP Institute Co., Ltd.
PHP研究所
1985

337
Kyoto-Fu Beikoku Symbol Mark
京都府米穀 シンボルマーク

Art Director & Designer
Akisato Ueda
上田晃郷

Client
Kyoto-Fu Beikoku Co., Ltd.
京都府米穀株式会社
1984

334

335

336

337

333

338

339

340

341

338
Pip Fujimoto Symbol Mark
ピップフジモト シンボルマーク

Art Director
Katsumi Yutani
油谷勝海

Designer
Shigo Yamaguchi
山口至剛

Client
Pip Fujimoto Co., Ltd.
ピップフジモト株式会社
1984

339
P.8.P. Logotype
ピー・エイト・ピー ロゴタイプ

Art Director & Designer
Shigo Yamaguchi
山口至剛

Client
Nippon Phonogram Co., Ltd.
日本フォノグラム株式会社
1985

340
NEW DISC Logotype
NEW DISC ロゴタイプ

Art Director
Shigo Yamaguchi
山口至剛

Designer
Hisako Nagano
永野寿子

Client
Swing Journal Co., Ltd.
株式会社スイングジャーナル社
1983

341
Chameleon Logo-Mark
カメレオン ロゴマーク

Art Director
Tsuyoshi Fukuda
福田 毅

Designer
Shigo Yamaguchi
山口至剛

Client
Creative Studio Chameleon
クリエイティブ スタジオ カメレオン
1981

criss cross 342

ZAHA 343

Ecsaine Plaza Aoyama 344

office Miyachi 345

342
Criss Cross Logotype
クリス クロス ロゴタイプ
Art Director & Designer
Kenichi Samura
佐村憲一
Client
Marimurá Co., Ltd.
馬里邑協團株式会社
1986

343
Zaha Logotype
ザハ ロゴタイプ
Art Director & Designer
Kenichi Samura
佐村憲一
Client
Vendome International
株式会社ヴァンドーム・インターナショナル
1986

344
Ecsaine Plaza Aoyama Logo-Mark
エクセーヌ-プラザ・アオヤマ ロゴマーク
Art Director & Designer
Kenichi Samura
佐村憲一
Client
Toray Industries Ecsaine Plaza Inc.
東レエクセーヌプラザ株式会社
1983

345
Office Miyachi Symbol Mark
オフィス ミヤチ シンボルマーク
Art Director
Kenichi Samura
佐村憲一
Designer
Hiroo Maruyama
丸山弘雄
Client
Office Miyachi
オフィス ミヤチ
1985

346
Sano Studio Symbol Mark
サノスタジオ シンボルマーク

Art Director & Designer
Harumasa Misaki
三崎陽尹

Client
Sano Studio Co., Ltd.
株式会社サノスタジオ
1986

347
Tokyo Royal Philharmony Orchestra Symbol Mark
東京ロイヤルフィルハーモニーオーケストラ シンボルマーク

Art Director & Designer
Harumasa Misaki
三崎陽尹

Client
Tokyo Royal Philharmony Orchestra
東京ロイヤルフィルハーモニーオーケストラ
1985

348
Sun System Laboratory Symbol Mark
サンシステム技研 シンボルマーク

Art Director & Designer
Harumasa Misaki
三崎陽尹

Client
San System Laboratory Co., Ltd.
株式会社サンシステム技研
1984

346

347

348

349
Japan Graphic Designers Association Symbol Mark
日本グラフィックデザイナー協会
シンボルマーク

Art Director & Designer
Kazuo Kishimoto
岸本一夫

Client
Japan Graphic Designers Association Inc.
社団法人日本グラフィックデザイナー協会
1979

349

350
Nichirei Logo-Mark
ニチレイ ロゴマーク

Art Director
Keiji Iijima
飯島啓司

Designer
Toshihiro Nakajima
中島寿洋

Client
Nichirei Corporation
株式会社ニチレイ
1985

350

351
Space Communications Corporation Symbol Mark
宇宙通信 シンボルマーク

Art Director
Akira Ishikawa
石川 明

Designer
Itaru Tamura
田村 至

Client
Space Communications Corporation
宇宙通信株式会社
1986

351

352
FCG Symbol Mark
フジサンケイグループ シンボルマーク

Art Director
Yuji Baba
馬場雄二

Designer
Katsu Yoshida
吉田カツ

Client
Fuji Sankei Communcations Group
フジサンケイコミュニケーションズグループ
1986

352

353
Tint Logotype
ティント ロゴタイプ

Art Director
Akira Inada
稲田 明

Designer
Akihiko Tsukamoto
塚本明彦

Client
Jurgen Lehl Co., Ltd.
株式会社ヨーガンレール
1980

353

354
Shiseido Nuda Logo-Mark
資生堂ヌーダ ロゴマーク

Art Director
Syunsaku Sugiura
杉浦俊作

Designer
Syuji Kawasaki
川崎修司

Client
Shiseido Cosmetics Co., Ltd.
株式会社資生堂
1984

354

356

357

358

355
By Zero Logo-Mark
By Zero ロゴマーク

Art Director
Katsumi Komagata
駒形克己
Koji Yanagisawa
柳沢光二

Designer
Katsumi Komagata
駒形克己

Client
J. Osawa & Co., Ltd.
株式会社大沢商会
1986

355

356
Shinfuji Paper Symbol Mark
新富士製紙 シンボルマーク

Producer
Honshupaper Co., Ltd. Packaging Center
本州製紙PC室

Art Director & Designer
Hidemichi Yamao
山尾日出道

Client
Shinfuji Paper Co., Ltd.
新富士製紙株式会社
1984

357
Mizuno Symbol Mark
水野本社 シンボルマーク

Art Director
Shunji Niinomi
新家春二

Designer
Yoko Sumi
鷲見容子

Client
Mizuno Co., Ltd.
株式会社水野本社
1982

358
Narita Golf Conrse Symbol Mark
成田ゴルフコース シンボルマーク

Art Director & Designer
Syoji Noguchi
野口正治

Client
Mitsuno Corporation
ミツノ株式会社
1980

359

360

361

362

363

364

359
Kuniumi Festival Symbol Mark
くにうみの祭典 シンボルマーク
Art Director & Designer
Motoo Hosino
星野元郎
Client
The Executive Committee of Kuniumi Festival, Hyogo Prefecture.
兵庫県くにうみの祭典実行委員会
1985

360
Vandai Symbol Mark
萬代 シンボルマーク
Art Director & Designer
Shigeyuki Hattori
服部成行
Designer
Makiko Hisamatsu
久松万紀子
Client
Vandai Co., Ltd.
株式会社萬代
1983

361
FM Sanin Symbol Mark
FM山陰 シンボルマーク
Art Director & Designer
Tadamasa Katsube
勝部忠正
Client
Fm Sanin Broadcasting Co., Ltd.
株式会社エフエム山陰
1986

362
EXPO '87 Tohoku Symbol Mark
未来の東北博覧会 シンボルマーク
Designer
Nobuya Nagata
永田暢也
Client
The Future of Tohoku Expo Associalion
未来の東北博覧会協会
1984

363
Sogo Palace Symbol Mark
ソーゴパレス シンボルマーク
Art Director & Designer
Hidemichi Yamao
山尾日出道
Client
Hotel Sogo Palace
ホテル ソーゴパレス
1982

364
Tosco Symbol Mark
トスコ シンボルマーク
Art Director & Designer
Kazuyoshi Miyata
宮田一義
Client
Tosco Co., Ltd.
トスコ株式会社
1985

365
School of Yashima Symbol Mark
ヤシマ経理専門学校 シンボルマーク

Consulting
MEDIART
株式会社メディアート

Art Director & Designer
Yoshiho Nakata
中田吉穂

Designer
N Design
N デザイン

Client
School of Yashima
学校法人ヤシマ情報経理専門学校
大阪校、鳳校
1986

365

366
J. Progressive Logo-Mark
J. プログレッスィヴ ロゴマーク

Art Director
Satoru Miyata
宮田 識

Designer
Satoe Inoue
井上里枝
Yumiko Nisimura
西村由美子

Client
J. Progressive Ltd.
株式会社J. プログレッスィヴ
1983

J. PROGRESSIVE LTD. 366

367
Axle International Symbol Mark
アクスル・インターナショナル シンボルマーク

Art Director & Designer
Shigeyuki Hattori
服部成行

Designer
Makiko Hisamatsu
久松万紀子

Client
Axle International Co., Ltd.
株式会社アクスル・インターナショナル
1985

367

368
Mandom Symbol Mark
マンダム シンボルマーク

Art Director
Yohei Akashi
明石洋平

Designer
Jam Creative Inc.
ジャム クリエイティブインコーポレイテッド

Client
Mandom Corp.
株式会社マンダム
1983

368

369
Vangard Symbol Mark
バンガード シンボルマーク

Art Director & Designer
Tatsuomi Majima
間嶋龍臣

Client
Vangard abvertising Co., Ltd.
株式会社バンガード・アドバタイジング
1981

369

370
Pittori Piccoli Symbol Mark
ピトリ・ピコリ シンボルマーク

Art Director & Designer
Sadao Sugaya
菅谷貞雄

Client
Pittori Piccoli, Inc.
株式会社ピトリ・ピコリ
1976

370

371

372

373

374

375

376

371
International Communications' Symbol Mark
国際コミュニケーションズ シンボルマーク

Art Director & Designer
Sadao Sugaya
菅谷貞雄

Client
International Communications' Incorporated
株式会社国際コミュニケーションズ
1980

372
Hotel Ocean Terrace Symbol Mark
ホテルオーシャンテラス シンボルマーク

Art Director
Takayuki Kuribayashi
栗林孝之

Designer
Kumi Shirahama
白浜久美

Creative Director
Ichiro Shimada
島田一郎

Client
HOTEL OCEAN TERRACE
ホテルオーシャンテラス
1986

373
Yubi Symbol Mark
ゆう美 シンボルマーク

Art Director & Designer
Takusei Tada
多田拓生

Client
Yubi Co., Ltd.
株式会社ゆう美
1982

374
Skynet Symbol Mark
スカイネット シンボルマーク

Designer
Tadanori Itakura
板倉忠則

Client
Skynet Communications Co., Ltd.
株式会社スカイネットコミュニケーションズ
1986

375
Sanko Symbol Mark
三紅 シンボルマーク

Designer
Tadanori Itakura
板倉忠則

Client
Sanko Co., Ltd.
株式会社三紅
1983

376
Melbourne/Osaka Double-handed Yacht Race Symbol Mark
メルボルン/大阪ダブルハンドヨットレース シンボルマーク

Art Director & Designer
Hirofumi Kine
木根博文

Client
Yamaha Osaka Cup '87 Promotion Council
Yamaha Osaka Cup '87 推進協議会
1986

377
Belle Mariee Logo-Mark
ベルマリエ ロゴマーク

Art Director & Designer
Yukiyoshi Sano
佐野幸義

Client
Suzuya Co., Ltd.
株式会社鈴屋
1978

378
Shikoku Taoru Kogyo Kumiai Symbol Mark
四国タオル工業組合 シンボルマーク

Art Director & Designer
Takahiro Shima
嶋　高宏

Client
Shikoku Taoru Kogyo Kumiai
四国タオル工業組合
1981

379
Copy-Writer Symbol Mark
コピーライター シンボルマーク

Art Director & Designer
Tathuomi Majima
間嶋龍臣

Client
Junichi Yamazaki
山崎淳一
1983

380
Hotel Niigata Symbol Mark
ホテル新潟 シンボルマーク

Art Director & Designer
Kiyokazu Watanabe
渡辺清和

Designer
Mika Saito
斎藤美佳

Client
Hotel Niigata Co., Ltd.
株式会社ホテル新潟
1986

381
Kisoken Logo-Mark
教育基礎研究所 ロゴマーク

Art Director & Designer
Sunao Kishi
岸　素直

Client
Kisoken
株式会社教育基礎研究所
1982

382
Fashion Town · Sapporo Symbol Mark
ファッションタウン・さっぽろ シンボルマーク

Designer
Yoshikatsu Tami
民　芳勝

Client
Fashion Town · Sapporo Promotion Council.
ファッションタウン・さっぽろ推進協議会
1984

383
Kakiuchi Corporate-Mark
カキウチ コーポレートマーク

Art Director & Designer
Tetsuo Simamura
島村哲夫

Client
Kakiuchi Co., Ltd.
カキウチ株式会社
1987

377

378

379

380

381

382

383

384
Society Logotype
ソサエティ ロゴタイプ

Art Director
Takatoshi Eguchi
江口隆敏

Art Director & Designer
Teruaki Hirose
広瀬輝明

Client
Noevir Co., Ltd.
株式会社ノエビア
1985

SOCIETY 384

385
Rim Logo-Mark
リム ロゴマーク

Designer
Yuji Dairokuno
大六野雄二

Client
Rikuyo-Sha Inter-Media, Inc.
株式会社リム［六耀社インターメディア］
1987

RIKUYO-SHA INTER-MEDIA 385

386
Bonton Logotype
ボントン ロゴタイプ

Art Director & Designer
Ken Miki
三木　健

Client
Maruhon Inc. Ito & Associates Div.
株式会社マルホン イトウ&アソシエイツ
1986

387
ZOOM Logotype
ZOOM ロゴタイプ

Art Director & Designer
Ken Miki
三木　健

Client
Maruhon Inc. Ito & Associates Div.
株式会社マルホン イトウ&アソシエイツ
1984

388
Kansai Logo-Mark
Kansai ロゴマーク

Art Director & Designer
Kyoji Nakatani
中谷匡児

Client
HI-SENSE Co., Ltd.
株式会社ハイセンス
1986

389
Isehara Logo-Mark
伊勢原 ロゴマーク

Designer
Mitsuo Hosokawa
細川光夫

Client
Isehara-City
伊勢原市
1986

390
BOMB! Logotype
ボム ロゴタイプ

Art Director & Designer
Kyoji Nakatani
中谷匡児

Client
Gakken Co., Ltd.
株式会社学習研究社
1981

391
Nimura Bankin Kogyo Logotype
二村板金工業 ロゴタイプ

Designer
Kazuharu Fujii
藤井和晴

Client
Nimura Bankin Kogyo Co., Ltd.
二村板金工業有限会社
1981

392
Gold Kowa Logo-Mark
ゴールドコーワ ロゴマーク

Art Director & Designer
Kyoji Nakatani
中谷匡児

Client
Gold Kowa
ゴールドコーワ
1986

386

387

388

389

390

391

392

393

394

395

396

393
Wakamathu Shinyokinko Symbol Mark
若松信用金庫 シンボルマーク

Art Director
Kazumasa Doi
土井一正

Designer
Hideo Toyomasu
豊増秀男

Client
Wakamathu Shinyokinko
若松信用金庫
1985

394
Kumamoto Technopolis Symbol Mark
熊本テクノポリス シンボルマーク

Art Director & Designer
Hideo Toyomasu
豊増秀男

Client
Kumamoto Technopolis Foundation
財団法人熊本テクノポリス財団
1985

395
Advisory Specialist for Consumers, Affairs" Symbol Mark
消費者生活アドバイザー シンボルマーク

Art Director & Designer
Hideo Toyomasu
豊増秀男

Client
The Japan Industrial Association
財団法人日本産業協会
1981

396
Fukuoka Fermer's Co-op. Symbol Mark
福岡市農業協同組合 シンボルマーク

Art Director
Asami Kishi
岸　潮美

Designer
Hideo Toyomasu
豊増秀男

Client
Fukuoka Fermer's Co-op.
福岡市農業協同組合
1986

397

398

399

400

401

397
Equal Employment Oppor-tunity Law Symbol Mark
男女雇用機会均等法 シンボルマーク

Art Director & Designer
Kimitake Yamamoto
山本圭武

Client
Policy Planning Divis'ion
労働省婦人局婦人政策課
1986

398
Kyowa Tsusho Symbol Mark
協和通商 シンボルマーク

Art Director & Designer
Kimitake Yamamoto
山本圭武

Client
Kyowa Tsusho Co., Ltd.
株式会社協和通商
1985

399
Paulista Phototype'-set-ting Labo" Symbol Mark
パウリスタ写植工房 シンボルマーク

Art Director & Designer
Kimitake Yamamoto
山本圭武

Client
Paulista Phototype'-setting Labo
パウリスタ写植工房
1983

400
Hanatsuru Symbol Mark
花ツル シンボルマーク

Art Director & Designer
Kimitake Yamamoto
山本圭武

Client
Hanatsuru Ltd. Responsi-bility Co.
有限会社フローリスト花ツル
1984

401
Bio Symbol Mark
バイオ シンボルマーク

Art Director & Designer
Kimitake Yamamoto
山本圭武

Client
Bio
バイオ
1984

402

403

404

405

402
Tashiro Electro Plating Engineers Symbol Mark
田代電化工業 シンボルマーク

Designer
Akiteru Nakajima
中島安貴輝

Client
Tashiro Electro Plating Engineers Co., Ltd.
田代電化工業株式会社
1979

403
Houtoku Hanbai Symbol Mark
ホウトク販売 シンボルマーク

Art Director
Shunji Niinomi
新家春二

Designer
Masatoshi Hirao
平尾正俊

Client
Houtoku Hanbai Co., Ltd.
ホウトク販売株式会社
1985

404
Yamao Design Office Symbol Mark
山尾デザイン事務所 シンボルマーク

Art Director & Designer
Hidemichi Yamao
山尾日出道

Client
Yamao Design Office
山尾デザイン事務所
1983

405
Fresser Symbol Mark
フレッサー シンボルマーク

Art Director & Designer
Hidemichi Yamao
山尾日出道

Client
Fresser Co., Ltd.
フレッサー株式会社
1982

406

407

408

409

410

411

406
Inbe Symbol Mark
いんべ シンボルマーク

Art Director & Designer
Yoshitake Sugimoto
杉本吉武

Client
Inbe Cleaning Co., Ltd.
有限会社いんべクリーニング店
1986

407
Japan A. V. L. Symbol Mark
日本A. V. L. シンボルマーク

Art Director & Designer
Masayoshi Tsuchiya
土屋昌義

Client
Japan A. V. L. Co., Ltd.
日本A. V. L. 株式会社
1984

408
Hiroshima City A. C. A. Symbol Mark
広島市農業協同組合 シンボルマーク

Creative Director
Ken Miki
三木 建

Art Director & Designer
Kris Markelin
クリス マークリン

Client
Hiroshima City A. C. A.
広島市農業協同組合
1986

409
TBR Symbol Mark
豊橋組紐ロープ工業 シンボルマーク

Art Director
Shunji Niinomi
新家春二

Designer
Yoko Sumi
鷲見容子

Client
TBR Co., Ltd.
豊橋組紐ロープ工業株式会社
1986

410
Deluta Symbol Mark
デルタ シンボルマーク

Conslting
Dentsu Inc.
株式会社 電通

Art Director
Takayuki Sashida
指田隆行

Designer
Yoji Yamamoto
山本洋司

Client
Toshiba Electronics Co., Ltd.
東芝電算機OAシステム事業部
1984

411
TOYOTA Logo-Mark
トヨタ ロゴマーク

Conslting
Nippon Design Center, Inc.
株式会社日本デザインセンター

Art Director & Designer
Yoji Yamamoto
山本洋司

Client
TOYOTA MOTOR CORPORATION
トヨタ自動車株式会社海外企画部
1985

412

413

414

415

416

417

412
Tokyo Shokuhin Symbol Mark
東京食品 シンボルマーク

Art Director & Designer
Yukio Kanise
蟹瀬行雄

Client
Tokyo Shokuhin Co., Ltd.
東京食品株式会社
1986

413
Mitsuboshi Symbol Mark
ミツボシ シンボルマーク

Art Director & Designer
Yukio Kanise
蟹瀬行雄

Client
Mitsuboshi Co., Ltd.
株式会社ミツボシ
1984

414
Bonboa Symbol Mark
ボンボア シンボルマーク

Art Director
Kazu Hishida
菱田 和

Designer
Yukio Kanise
蟹瀬行雄

Client
Restaurant Bonboa
レストランボンボア
1976

415
JAFFE Symbol Mark
ジャッフェ シンボルマーク

Art Director & Designer
Ken Miki
三木 健

Client
Japan Food Engineering Co., Ltd.
株式会社ジャパン・フード・エンジニアリング
1987

416
Cawaju Symbol Mark
かわ重 シンボルマーク

Art Director & Designer
Kigen Oki
大木紀元

Client
Cawaju
かわ重
1985

417
National Capital Area Natural Footway Symbol Mark
首都圏自然歩道 シンボルマーク

Art Director & Designer
Hidemichi Yamao
山尾日出道

Client
National Capital Area Natural Footway Liaison Committee
首都圏自然歩道連絡協議会
1981

418
Modepia Symbol Mark
モードピア シンボルマーク

Art Director & Designer
Isao Kimura
木村勲生

Client
All Style Co., Ltd.
オールスタイル株式会社
1986

418

419
GOÛTER DE MAMAN Logo-Mark
グーテ・ド・ママン ロゴマーク

Art Director
Kazuo Shimada
嶋田和夫

Designer
Keiko Mitomi
三富恵子

Client
GOÛTER DE MAMAN
グーテ・ド・ママン
1985

419

420
The Hop Symbol Mark
ザ・ハップ シンボルマーク

Art Director
Yohei Akashi
明石洋平

Designer
Jam Creative Inc.
ジャム クリエイティブ インコーポレイテッド

Client
Keihan Electric Rail-Way Co., Ltd.
京阪電気鉄道株式会社
1985

421
ROBUNDO Logotype
朗文堂 ロゴタイプ

Art Director & Designer
Yoshihiro Yoshida
吉田佳広

Client
Robundo, Inc.
株式会社朗文堂
1983

422
Fifth Logo-Mark
フィフス ロゴマーク

Creative Director
Yoshihiro Iwanaga
岩永嘉弘

Art Director
Iwao Matsura
松浦厳生

Designer
Kenji Takumi
託見健二

Client
Shinjuku Tokyo Kaikan Co., Ltd.
サントリー新宿東京会館株式会社
1983

423
ACT-1 Logotype
アクトワン ロゴタイプ

Art Director
Shunji Niinomi
新家春二

Designer
Yoko Sumi
鷲見容子

Client
ACT-1 Co., Ltd.
アクトワン株式会社
1978

A Mecca for fine fashions, the discriminate are assured their clothing prayers answered.

420

421

422

423

424
Styling Logotype
スタイリング ロゴタイプ
Creative Director
Ichiro Shimada
島田一郎
Art Director
Takayuki Kuribayashi
栗林孝之
Designer
Kumi Shirahama
白浜久美
Client
Styling International Inc.
株式会社スタイリングインターナショナル
1985

425
Nissan Motor Logotype
日産自動車 ロゴタイプ
Conslting
Ishiwatari Hisao Office
石渡日佐夫企画事業所
Art Director
Nissan Motor
日産広報室
Type Director
Yasaburo Kuwayama
桑山弥三郎
Designer
Ikuo Masubuchi
増渕育夫
Client
Nissan Motor Co., Ltd.
日産自動車株式会社
1985

426
Kasugai Logotype
カスガイ ロゴタイプ
Art Director
Syozo Murase
村瀬省三
Designer
Jun Murase
村瀬 潤
Client
Kasugai Co., Ltd.
カスガイ株式会社
1984

427
Societe Fan Fan Logotype
ソシェット ファンファン ロゴタイプ
Art Director & Designer
Koji Iwagami
岩上孝二
Client
Societe Fan Fan
ソシェット ファン ファン
1983

428
Cash Logo-Mark
キャッシュ ロゴマーク
Art Director & Designer
Koji Iwagami
岩上孝二
Client
Cash
キャッシュ
1982

STYLING 424

マーチ ローレル 425

KASUGAI 426

FAN FAN 427

CASH
DESIGNERS MEMBERSHIP CLUB 428

429

430

431

432

429
Gifu Chubu Future Exposition Symbol Mark
ぎふ中部未来博覧会 シンボルマーク

Art Director & Designer
Fumio Koyoda
小代田文男

Client
Gifu Chubu Future Exposition Association
財団法人ぎふ中部未来博覧会協会
1986

430
Kobe Port Island Exposition Symbol Mark
神戸ポートアイランド博覧会 シンボルマーク

Creative Director
Takeshi Otaka
大高 猛

Art Director & Designer
Fumio Koyoda
小代田文男

Client
Kobe Port Island Exposition Association
財団法人神戸ポートアイランド博覧会協会
1979

431
Aki Design Room Symbol Mark
瞕デザイン室 シンボルマーク

Art Director & Designer
Fumio Koyoda
小代田文男

Client
Aki Design Room
瞕デザイン室
1984

432
International Exhibition Center, Osaka Symbol Mark
大阪国際見本市会場 シンボルマーク

Art Director & Designer
Fumio Koyoda
小代田文男

Client
Osaka City・Osaka International Business Promotion Center
大阪市・財団法人大阪国際経済振興センター
1985

433

434

435

436

437

438

433
Taiheiyo Marine Symbol Mark
太平洋海事 シンボルマーク
Art Director & Designer
Koichi Watanabe
渡部孝一
Client
Taiheiyo Marine Co., Ltd.
太平洋海事株式会社
1981

434
Yoshino Interior Symbol Mark
ヨシノインテリア シンボルマーク
Art Director
Koichi Watanabe
渡部孝一
Designer
Kunio Watanabe
渡部邦雄
Client
Yoshino Interior Co., Ltd.
株式会社ヨシノインテリア
1976

435
Ota Dental Clinic Symbol Mark
オオタ歯科医院 シンボルマーク
Art Director & Designer
Koichi Watanabe
渡部孝一
Client
Ota Dental Clinic
オオタ歯科医院
1974

436
Marakai Corporation Symbol Mark
マルカイ コーポレーション シンボルマーク
Art Director & Designer
Koichi Watanabe
渡部孝一
Client
Marukai Corporation
株式会社マルカイコーポレーション
1981

437
Central Marine Symbol Mark
セントラルマリン シンボルマーク
Art Director & Designer
Koichi Watanabe
渡部孝一
Client
Central Marine Co., Ltd.
セントラルマリン株式会社
1984

438
Tate Symbol Mark
テイト シンボルマーク
Art Director & Designer
Koichi Watanabe
渡部孝一
Client
Tate Co., Ltd.
株式会社テイト
1982

439
Taishin International Symbol Mark
タイシンインターナショナル シンボルマーク

Creative Director
Shinpei Sai
佐井新平

Art Director & Symbol Mark Designer
Yoneo Jinbo
神保米雄

Logotype Designer
Chizuko Toda
戸田智津子

Client
Taishin International Co., Ltd.
株式会社タイシンインターナショナル
1985

439

440
MU Logotype
ムー ロゴタイプ

Art Director & Designer
Kyoji Nakatani
中谷匡児

Client
Gakken Co., Ltd.
株式会社学習研究社
1982

440

441
KYOCERA Symbol Mark
京セラ シンボルマーク

Art Director & Designer
Mitsuo Hosokawa
細川光夫

Client
Kyocera Corporation
京セラ
1983

441

442
Kumon Logo-Mark
公文 ロゴマーク

Art Director & Designer
Mitsuo Hosokawa
細川光夫

Client
KUMON INSTLTUTE OF EDU-CATION
株式会社公文教育研究会
1983

443
Hoshiden Symbol Mark
星電器 シンボルマーク

Art Director
Ken Miki
三木 建

Designer
Akihiko Tamura
田村昭彦

Client
Hoshiden Electronics Co., Ltd.
星電器製造株式会社
1979

444
Yamada Foods Symbol Mark
ヤマダフーズ シンボルマーク

Art Director & Designer
Masahiko Muto
武藤正彦

Client
Yamada Foods Co., Ltd.
株式会社ヤマダフーズ
1986

445
Fashon Care Muromachi Symbol Mark
ファッションケア室町 ロゴマーク

Consulting
Hattori Concept Office Co., Ltd.
株式会社服部コンセプト事務所

Art Director
Yuzi Dairokuno
大六野雄二
Koichi Kurenuma
暮沼宏一

Designer
Yuzi Dairokuno
大六野雄二

Client
Tokushima Self Dry Joint System Co-operative.
徳島セルフドライ協業組合
1986

446
Partnership Symbol Mark
パートナーシップ シンボルマーク

Designer
Ryoichi Matsuda
松田良一

Client
Hokuriku Creators Partnership
北陸クリエイターズパートナーシップ
1981

447
Nakamoto Gankyo Symbol Mark
中本眼鏡 シンボルマーク

Designer
Ryoichi Matsuda
松田良一

Client
Nakamoto Gankyo Inc.
株式会社中本眼鏡
1986

MASAMI MATSUI 448

449

REEF CORPORATION 450

451

452

SAGAN 453

448
Masami Matsui Symbol Mark
松井雅美 シンボルマーク

Art Director
Hiroshi Takahara
高原 宏

Designer
Ichiro Sogo
十河一郎

Client
Axe Company Limted
株式会社アクス
1985

449
H・S Symbol Mark
西城秀樹 シンボルマーク

Art Director
Hiroshi Takahara
高原 宏

Designer
Hayahiko Yonaga
世永逸彦

Client
Earth Coporation
アース・コンポレーション
1984

450
Reef Logotype
リーフ ロゴタイプ

Art Director
Takanori Aiba
相羽高典

Designer
Hiroyuki Hayashi
林 廣行

Client
Reef Corporation
株式会社リーフ
1986

451
Oji Tape Symbol Mark
王子テープ シンボルマーク

Art Director
Yutaka Takama
高間 豊

Designer
Michiko Ozeki
大関美智子

Client
Oji Kako Co., Ltd.
王子化工株式会社
1977

452
non・no more books Symbol Mark
ノンノ・モア ブックス シンボルマーク

Art Director & Designer
Kyoji Nakatani
中谷匡児

Client
Shueisha Publishing Co., Ltd.
株式会社集英社
1982

453
Sagan Logo-Mark
サガン ロゴマーク

Art Director & Designer
Kyoji Nakatani
中谷匡児

Client
Sagan
サガン
1986

454
edic Logo-Mark
edic ロゴマーク

Consulting
Saburo Tomoto
東本三郎

Art Director & Designer
Kyoji Nakatani
中谷匡児

Client
edic
中央出版edic事業部
1986

455
Az Symbol Mark
エーゼット シンボルマーク

Art Director
Ken Miki
三木 建

Designer
Jam Creative Inc.
ジャム クリエイティブ インコーポレイテッド

Client
Az Co., Ltd.
エーゼット商事
1981

456
Mr. and Mrs. Yasui Symbol Mark
保井夫婦 シンボルマーク

Designer
Yutaka Sato
佐藤 豊

Client
Mr. Toshio Yasui and Mrs. Kimie
保井俊男・君江夫婦
1979

457
TPCE Symbol Mark
TPCE シンボルマーク

Art Director & Designer
Minoru Niijima
新島 実

Designer
Kenzo Kobayashi
小林健三

Client
Tokyo Paint Commerce Association
東京塗料商業協同組合
1984

458
Beruf Logo-Mark
ベルーフ ロゴマーク

Creative Director
Masaaki Izumiya
泉屋政昭

Art Director & Designer
Masami Shimizu
清水正己

Client
Recruit Co., Ltd.
株式会社リクルート
1986

459
Nutty House Logo-Mark
ナッティーハウス ロゴマーク

Art Director
Nobuyuki Yasuda
安田信行

Designer
Satomi Fuziwara
藤原さとみ

Client
TOYO NUT Co., Ltd.
東洋ナッツ食品株式会社
1981

460
Alcione Logo-Mark
アルチオネ ロゴマーク

Art Director
Nobuyuki Yasuda
安田信行

Designer
Satomi Fuziwara
藤原さとみ

Client
Alcione Co., Ltd.
株式会社アルチオネ
1982

461
Natural Foods Logo-Mark
ナチュラルフーズ ロゴマーク

Art Director
Nobuyuki Yasuda
安田信行

Designer
Satomi Fuziwara
藤原さとみ

Client
TOYO NUT Co., Ltd.
東洋ナッツ食品株式会社
1981

462
Hotel Hiltop Logo-Mark
ホテルヒルトップ ロゴマーク

Art Director
Yoichi Tanoue
田上洋一

Designer
People AD.
あど・ぴいぷる

Client
Yuge Co., Ltd.
株式会社ユゲ
1981

463
Animal Care Logo-Mark
アニマルケア ロゴマーク

Art Director & Designer
Tatsumi Kuwahara
桑原達美

Client
Animal Care Co., Ltd.
株式会社アニマルケア
1980

464
Hanakyabetsu Symbol Mark
花きゃべつ シンボルマーク

Art Director
Shunji Niinomi
新家春二

Designer
Yoko Sumi
鷲見容子

Client
Tea Room Hanakyabetsu
Tea Room 花きゃべつ
1978

458

459

460

461

462

463

464

465
Rhino Logo-Mark
ライノ ロゴマーク
Art Director
Yukiyoshi Sano
佐野幸義
Designer
Yoshihisa Syouji
荘司芳久
Client
World Co., Ltd.
株式会社ワールド
1982

466
Sweet Temptation Logo-Mark
スウート・テンプテーション ロゴマーク
Art Director
Yukiyoshi Sano
佐野幸義
Designer
Toshimityu Ohkubo
大久保寿光
Client
Lofty Co., Ltd.
ロフティー株式会社
1987

467
Sistiene Symbol Mark
システィーヌ シンボルマーク
Art Director
Syozo Murase
村瀬省三
Designer
Jun Yosida
吉田 純
Motokazu Uchida
内田元和
Client
Sistiene Co., Ltd.
システィーヌ株式会社
1984

468
Hang-Glider Logo-Mark
ハングライダー ロゴマーク
Art Director
Yonefusa Yamada
山田米英
Designer
Sin Sasaki
佐々木 伸
Client
Yamada Design Room
山田デザイン室
1984

469
Vigla Symbol Mark
ヴィグラ シンボルマーク
Designer
Akira Ishikawa
石川 明
Reiko Takahashi
高橋玲子
Client
JAPAN TOBACCO INC.
日本たばこ産業株式会社
1986

470
Shintaku Obstetrics & Gynecology Hospital Symbol Mark
新宅産婦人科病院 シンボルマーク
Art Director & Symbol Mark Designer
Hiromitsu Kono
河野博光
Assistant Designer
Yasuyuki Koga
古賀保行
Client
Shintaku Obstetrics & Gynecology Hospital
新宅産婦人科病院
1986

471
Natural House Symbol Mark
ナチュラルハウス シンボルマーク
Art Director & Designer
Hiromitsu Kono
河野博光
Client
Natural House Corporation
株式会社ナチュラルハウス
1980

472
Asahi Annai Mark
朝日案内 マーク
Designer
Keijiro Ozumi
小住慶二郎
Client
Asahi Shimbun Publishing Company
朝日新聞西部本社
1960